AT THE EDGE OF UNCERTAINTY

MICHAEL BROOKS

11 Discoveries Taking
Science by Surprise

P

PROFILE BOOKS

First published in Great Britain in 2014 by
PROFILE BOOKS LTD
3A Exmouth House
Pine Street
London EC1R 0JH
www.profilebooks.com

10 9 8 7 6 5 4 3 2 1

Printed and bound in Great Britain by Clays, Bungay, Suffolk

A CIP catalogue record for this book is available from the
British Library.

ISBN 978 1 78125 127 0
eISBN 978 1 84765 966 8

The paper this book is printed on is certified by the © 1996 Forest
Stewardship Council A.C. (FSC). It is ancient-forest friendly. The printer
holds FSC chain of custody SGS-COC-2061

FSC
www.fsc.org
MIX
Paper from
responsible sources
FSC® C018072

Michael Brooks is the author of the bestselling non-fiction title *13 Things That Don't Make Sense* and *Free Radicals: The Secret Anarchy of Science*. He holds a PhD in quantum physics, is a consultant at *New Scientist* and writes a weekly column for the *New Statesman*.

Also by Michael Brooks

13 Things That Don't Make Sense
Free Radicals

Knowledge is personal and responsible, an unending adventure at the edge of uncertainty.

Jacob Bronowski

CONTENTS

ACKNOWLEDGEMENTS

A project like this always involves battling with uncertainty, and I am grateful to many people for helping to shape and inform the content of this book. The enthusiasm of Andrew Franklin and his colleagues at Profile provided steadfast support through the process of pulling these ideas and pages together. My agent, Caroline Dawnay at United Agents, has been an ever-present source of helpful advice and direction. I am indebted to the many people who responded to my pestering emails, giving freely of their time to read through various parts of the manuscript and point out where there was room for improvement. Particularly worthy of thanks are Giovanella Baggio, Martin Bobrow, Daniel Bor, Carl Gibson, Neal Halsey, Grazyna Jasienska, Rajeev Krishnadas, Karen Lillycrop, Toby Ord, Krill Rossiianov, Rudolph Schild, Jack Copeland, Hava Siegelmann and Vlatko Vedral. They helped limit my uncertainties, but any remaining errors are mine.

As ever, I must thank those closest to me for coping with more months in the shadow of my inattention. Phillippa, Millie and Zachary, our family is a deep well of certainty; thank you.

INTRODUCTION

*Daring ideas are like chessmen moved forward. They may be
beaten, but they may start a winning game.*

Johann Wolfgang von Goethe

You might think it's hard to take science by surprise. After all,
aren't scientists the clever ones, the know-it-alls? Aren't they
revered as the people with answers to every question?

It's certainly true that science has made extraordinary inroads
into discovering how the universe and everything within it ticks
along. Science has been successful for the most part in explaining why things are as they are. But in the process they have also
discovered the broad horizon of their ignorance.

That is not a problem; on the contrary, it is an enormous gain.
In science, ignorance is not something to be ashamed of, something to hide, but something to acknowledge and explore. Just as
the tide's ebb and flow created the perfect conditions for life to
arise at the edge of Earth's oceans, the place where certainty gives
way to uncertainty – the shoreline of our ignorance – is fertile
ground indeed.

In much of science, the parts we know well, there is relatively little to be gained. Here, further up the beach, we might

determine a constant to another decimal place; there, we seek to make a slightly more accurate measurement of the time it takes for a signal to travel between neurons in the brain. We find a catalyst that will make a chemical reaction happen a little more quickly or efficiently. We discover another distant star to enter into our catalogues, and so on. Such incremental gains are always there for the taking, pebbles to be turned over and inspected. These advances are added to the canon of science, but they don't change anything – not really. That is why they don't make front-page news. Newton was too humble when he wrote about his life's work shortly before his death. He said, 'I was like a boy playing on the sea-shore, and diverting myself now and then finding a smoother pebble or a prettier shell than ordinary, whilst the great ocean of truth lay all undiscovered before me.' It's not true: much of what he did was reaching into the murky water and pulling out surprising new truths.

Many have followed in his footsteps, moving out of the safe zone, venturing beyond the very limits of our knowledge and peering into the gloom until they could make out the vague shape of something intriguing. Then, grabbing all the tools at their disposal, they plunged into the water, intent on bringing that shadowy form back on to dry land.

It is a dangerous thing to do. Here at the edge of uncertainty we have discovered shocking things – things that have made some scientists beat a hasty retreat. It was here, for instance, that Henri Poincaré discovered that a resolution to certain anomalies in electromagnetic theory would require rethinking the nature of time. Poincaré was too perturbed by the discovery to press on; it was left to Albert Einstein to venture into dark waters and hunt out the special theory of relativity. The astronomer Arthur Eddington had once done some work that suggested the existence of black holes, but he hated the implications: that there were rips in the fabric of the universe. So when Subrahmanyan Chandrasekhar confirmed the suggestion with a mathematical proof, Eddington railed against it, and made Chandrasekhar's life a misery. Neuroscientist Benjamin Libet was another fugitive from unwelcome truth; when he performed an experiment showing that humans lacked free will, he dedicated the rest of his life to proving himself wrong. Good science – important science – can be as unnerving as it is enlightening.

Sometimes work at the edge of uncertainty is without tangible fruit: it simply uncovers our ignorance. From time to time, for example, we will discover that our previous scientific understanding was built on a flimsy foundation and must be urgently shored up – or even abandoned. This is not the disaster it might seem, because science is fickle: it reserves the right to change its mind. Some scientists might make definitive statements, but others must then take on the task of trying to undermine them. Very often they succeed: new experiments, new thoughts and new discoveries turn our thinking on its head, reverse a trend, expose the flaws in previous experiments, or poke holes in a celebrated scientist's thinking. The initial result is usually panic or denial, anger or derision – often all of the above. Eventually, though, after months, a year, a decade or a century, there is resigned acceptance of the new. Until, that is, someone dares to take in the view from the new edge of uncertainty. That novel perspective inevitably leads to further revisions and revolution. 'Everything we know is only some kind of approximation,' Richard Feynman once said. 'Therefore, things must be learned only to be unlearned again or, more likely, to be corrected.' This is where Galileo, Newton, Darwin and Einstein did their work. All the revolutionaries have been challenged, accepted, then challenged again. As George Bernard Shaw put it, 'All great truths begin as blasphemies.'

Where science does have a problem is in the fact that our

collective memories are so short. Once that resigned acceptance of a discovery comes, we forget that there was once such a kerfuffle. We act as if this truth were always with us, that it is self-evident. We forget the decades of persecution someone endured in order to shepherd us to the view we would now die to defend. And so we become comfortable – so comfortable that we will wantonly persecute the man or woman who comes to disturb our peaceful state. Take the atom, for example. No one now denies its existence, and it seems impossible that anyone thought it a pointless fiction. The atom is part of our world-view, part of our language, part of our collective history. But it wasn't always thus, as the tragic story of the Austrian physicist Ludwig Boltzmann shows so clearly.

These days Boltzmann would almost certainly be diagnosed as having bipolar disorder. His moods swung between elation and deep depression. When up, he was convivial – his students loved him and lectures at the University of Vienna were sometimes attended by so many people that the overspill ran into the corridor and down the stairs from the lecture hall. When down, a phase often triggered by the rejection of his peers, his moods were very dark indeed. In 1900, for instance, after an argument with a member of his department, Boltzmann tried to take his own life.

The rancour was always over the existence of atoms. Boltzmann was convinced that they existed in some form or other; most of his colleagues, some of the most powerful men of science at the time, were convinced they did not. Though the notion of atoms seems rather obvious to us now, and might well have seemed obvious to physicists hundreds of years earlier, many of Boltzmann's contemporaries were obsessed by the nebulous concept of energy. The industrial revolution had raised energy, in their minds, to a position where it became the fundamental component of reality. They believed that the new science of thermodynamics, which had been constructed to further the gains of the industrial revolution from its roaring heat engines, provided reality's rules.

Boltzmann spent the latter part of his working life countering this view. He constructed intricate arguments which proved that the mechanical motion of atoms was the fundamental driving force of gases as they heated, expanded, cooled and contracted. The theory was statistical, not absolute: though individual atoms followed simple rules, together they would create a variety of observable outcomes. Some outcomes were more likely – some much more likely – than others, providing an explanation for the observed phenomena. It was an unpopular notion, with the most popular physicists of the time railing against it. Chief in opposition was Ernst Mach, who admitted that atoms could be a useful crutch for thinking about reality but nothing more: the atom, he said, 'must remain a tool for representing phenomena'.

Boltzmann defended his position with gusto, but was undoubtedly worn down by the fights and his opponents' air of indifference. In the middle of one debate, he recalled, 'Mach spoke out from the group and laconically said: "I don't believe that atoms exist." This sentence went round and round in my head.'

Boltzmann's head had never been a terribly stable place, and the years of bitter feuds over the atom were a further strain. Eventually Boltzmann decided to end the agony for good. In 1906, while his wife and daughter swam nearby in the blue waters of the Bay of Duina near Trieste, Boltzmann hanged himself. His daughter, sent back to check on her father, found his body hanging by a short cord from a window casement. For the rest of her life, she never once spoke of what she had seen.

The singular literary character Lemony Snicket seems to understand the predicament Boltzmann found himself in. 'It is very unnerving to be proven wrong,' he says in *The Reptile Room*, 'particularly when you are really right and the person who is really wrong is proving you wrong and proving himself, wrongly, right.'

Such is the eternal dilemma of science: it's not always clear who is right, and the truth sometimes emerges too late for its champion to enjoy the victory. We don't know that Boltzmann's suicide resulted directly from the opprobrium raised against him, but we do know he played a vital – if tragic – role in securing a victory for a new and better understanding of reality. Within a few years of his death, observations of pollen and dust grains being knocked about at random by invisible entities led to the acceptance of Boltzmann's atom.

Historians of science know this well, but historical examples are of limited use. 'To imagine that turmoil is in the past and somehow we are now in a more stable time seems to be a psychological need,' the geologist Eldridge Moores once said. He was talking about wishful thinking concerning the stability of the ground beneath our feet, but he might have been talking about science. It's somehow easier to marvel at the fossils – to enjoy these stories of science's evolution – than to accept that things might still be evolving, that there is still an edge of uncertainty over which we can peer.

That is what this book has to offer: a glimpse of today's edge. We should be glad that there is one. After all, we don't have forever to mess about dotting the i's and crossing the t's of science. According to biologists John Lawton and Robert May, the fossil record tells us that a mammalian species lasts a million years on average. We have already been around for one-fifth of that time, and only recently started to study the world around us using what we now recognise as scientific techniques. Maybe the discoveries we make at the edge of uncertainty will help our species be the first to last forever – perhaps what we are studying here is the key to eternity. We certainly won't find the secrets of survival in the extra decimal place of something we know too well already.

We must go down the beach and peer into the dark waters. Extraordinary things are waiting to be discovered. They are almost certainly too extraordinary for us to cope with just yet. But we are becoming accustomed to the murky view, and a few hints and teases are beginning to resolve before our eyes. It is those hints and teases that we are about to explore; the chapters of this book describe some of the dangerous frontiers of science today. They are things we see most dimly, and yet feel most drawn to because of their potential to transform the way we see our selves and the way we live out our existence.

We start with an acknowledgement of what is perhaps our greatest scientific weakness: the human brain. These few pounds of wobbly biology are our only means of understanding the universe, and yet we don't even know what it means to 'understand'. We think, therefore (we tell ourselves) we are: we have our sense of self. From there we extrapolate to see our selves playing a role in a grand cosmic theatre.

In many ways it is a remarkable, self-aggrandising world-view. After all, as we will see, many other species of animal – species we don't see as struggling with existential questions – are remarkably similar to us. Our discoveries about the abilities of non-human animals has not knocked us off the pinnacle of creation, but it has brought many of our fellow creatures up to join us. These days we highlight the similarities, not the differences. One of the consequences is a movement to suggest we can merge human and non-human animals for medical purposes. It's something we are already doing with the creation of chimeras, though this is still at the edge of uncertainty and only slowly emerging from the scientific (and ethical) darkness.

We are also discovering that we have been transgressing a couple of other ethical boundaries. Were the geneticist Jacob Bronowski, the man who said that knowledge is 'personal and responsible', still with us today, he wouldn't have hesitated to suggest that our emerging understanding of epigenetics necessitates a headlong chase into the unknown. Epigenetics describes how genes work differently in our bodies after assault and insult from factors generally associated with poverty, deprivation and pollution. The effects are problematic and long-lasting, sometimes cascading down the generations. We are only just beginning to realise how personal biology is – and how responsible our response should be.

The same can be said of our discoveries about the role of gender in medicine: somehow we have been unconscionably crude in our medical approach to human beings. Did we really think that, the obvious points aside, gender makes no difference? We don't any more. Perhaps more forgivable, given the paucity of our understanding of the brain, was our neglect of the mind's power in the body. Here too, though, we are slowly undoing the cynicism of those who were happy to dress up their ignorance as knowledge and understanding.

If only we could do the same with those who tout quantum theory as the key to health. There is an undoubted appeal in calls to employ 'the power of quantum healing to transcend disease and aging', as the mystic Deepak Chopra would have us do. However, it is a mirage in the desert. The truth is, we are only just learning how to make the most tentative investigations of the role quantum physics plays in biology. It does seem that there are areas where nature has exploited the strange rules that govern atoms and molecules to create new opportunities for life to blossom and flourish in adversity. But here, at this boundary between life and the stuff of the cosmos, we truly jump into the deep waters.

As we bring together the sum of our experience in mathematics and physics, in experiment and theory, we can make the tentative suggestion that the universe is a computer, with our thoughts and actions acting as the programs whose instructions create what our brains (our poorly understood brains, remember) interpret as reality. Is this as delusional as Newton's insistence on the 'clockwork heavens' – an interpretation based on the technology of his time? Perhaps. After all, the computer is only a few decades old, and its inventor, Alan Turing, did see another kind of computing machine beyond the one that is familiar to us. Perhaps the hypercomputer will be a better guide to reality.

Not that we are finished with the reality we already know. There are those who would like to close the book on our story of how the cosmos came to be, but others are resisting. There are too many holes in the Big Bang story, and too many places where we have plugs that might fit – or at least know how we might fashion them to fit. It may be that, when we have patched up the history of the universe, there is more patch than fabric and we need to start again. We are already starting again with one of its fundamental constituents: the flow of time, it seems, is nothing more than an illusion. The tick-tock of passing moments is all in our minds, the physicists suggest.

In many ways, it would be easier to ignore all this, to go back up the beach to where the unfinished i's and t's are waiting. After all, we are simple creatures, easily fooled by our senses, our inner logic and our desire to bring simplicity to our interactions with the world. These difficult topics expose our weaknesses and leave us open to failure. Making sense of them is hard.

The beauty of human beings, though, is that we are fierce and indefatigable. We have shown ourselves determined to grapple with the universe around us until it surrenders its secrets to our inquiries. That is why we go to the edge of uncertainty: to quest, and question, and fight with ourselves and others until we have an answer. Then, aware that we have brushed against other questions and surprises, we stow our new discoveries safely, and dive back into the dark waters to wrestle more things into the light. We have been doing it for centuries, and we can only hope we will be doing it for centuries to come. This is, after all, the best thing humans have ever done.

This is how those mysterious and powerful brains compel us to behave: they endow us with the curiosity, the bravery and the tenacity to hunt out the truth as best we can. It's not an easy way to live. By the end of this journey to the frontiers of human certainty and beyond, your brain will feel battered and bruised. But it will also cry out for more. Adventuring is addictive. You have been warned.

1

TRIUMPH OF THE ZOMBIE KILLERS

The science of consciousness has
risen from the grave

*We have been to the moon, we have charted the depths of the ocean and
the heart of the atom, but we have a fear of looking inward to ourselves
because we sense that is where all the contradictions flow together.*

Terence McKenna

To the audience's utter delight, Gustav Kuhn is performing
magic tricks. He makes ping-pong balls disappear and reap-
pear in ridiculous places. Then he explains how he did it. 'It's
simple misdirection,' he says. 'I manipulate your attention by
moving my hand in certain directions; you can't help but follow
it with your eyes, which gives me the chance to...' He turns his
head, and our gaze follows. The ball is back in his hand. We can't
help but applaud.

It's unusual for an audience to be clapping this early into
a scientific talk. Usually there's a smattering of applause at
the end – often a manifestation of relief that it's finally over.
But here at the sixteenth meeting of the Association for the

Scientific Study of Consciousness the audience is enthralled from the start.

Kuhn thinks there should be a science of magic. The effects he and other magicians create are robust, significant, replicable and, above all, useful, he says – just like good scientific results, in other words. He and his co-presenter, Ronald Rensink, another magician–scientist, think that studying what magicians do can teach us about perception and cognition (and deception), how children develop an understanding of what is possible and impossible, why magical beliefs persist and what happens when the brain develops in unexpected ways. A study of magic could help us develop new tricks for engaging and interacting with people and technology and find new angles on problem-solving. And most important of all, it might give us a window on what it means to be conscious.

Studying consciousness used to be considered the ultimate waste of time. It is, after all, a subjective phenomenon, and thus unlike anything else in science. How can I study someone else's consciousness when I have to rely on their reports? How can I study my own, when I can't get any distance from it? Somehow, that spongy matter inside my skull creates something we call consciousness, but if I probe it, I disturb it. We don't have the means to keep a brain alive outside the skull and, even if we did, would we expect to dissect a brain and find its consciousness?

In 1994 philosopher David Chalmers coined a phrase about consciousness that has become a millstone or a mantra, depending on your point of view: 'The Hard Problem'. Consciousness 'escapes the net of reductive explanation,' Chalmers says. 'No explanation given wholly in physical terms can ever account for the emergence of conscious experience.' In other words, consciousness can't be explained by reverse-engineering the brain. You can't build a brain and expect to trace where its consciousness comes from. Consciousness is different in character from the set of all physical facts – it stands apart. That's why, he said, it is possible we are surrounded by undetectable zombies.

Any number of movies have described the onset of a zombie apocalypse. In not one of them has the hero used sleight-of-hand tricks to give their loved ones time to get away. That might seem like a banal observation, but it raises an interesting question about the nature of consciousness – and Chalmers's argument. Would a zombie be amazed and distracted by Kuhn's conjuring? What do zombies make of magic?

To be fair to Chalmers, he wasn't talking about the familiar, flesh-hanging-off, undead, food-obsessed zombies of science fiction. After all, they're easy to spot, with their lumbering gait, their insensitivity to pain or injury, their inability to communicate with or relate to others, and their glassy-eyed stare. What we're talking about is the perfect copy of a normal human, one that, from the outside, looks no different to you or me. This zombie walks normally; it can hold a conversation. It will even tell you it is feeling something. But the first question you have to ask yourself is how you could tell if it is telling the truth. You couldn't.

You can say exactly the same about your work colleagues. You, as Descartes pointed out, know you are conscious – '*cogito, ergo sum*', I think, therefore I am. But how do you know anyone else is conscious? All you have to go on is the fact that they appear to be the same as you. They react to stimuli such as a punch in the arm in the same way you would. Ask them a question and they respond in reasonable ways, and in a reasonable time. But if you ask them what they are experiencing, you have no way of knowing if they are just telling you what they think you expect them to say. They might not feel anything – they might just know what a human being is expected to be feeling in that situation, and report that.

This is the zombie hypothesis: that everyone around you might lack any self-awareness, any sense of self, and you wouldn't necessarily know it. Bringing it closer to home, imagine a version of you that is exactly like you, physically and mentally, so they look, act and speak like you, even thinking like you to give the same answers that you would to any question someone cared to pose. The difference is that this version of you has no awareness of themselves; they are, effectively, an automaton.

The fact that you can imagine this, Chalmers says, means it is theoretically possible. And so, he argues, consciousness must be something extra and above the physical material and processes of our brains, something that sits on top of our sensory perceptions, our reactions to them and our reporting of them.

That 'something' makes us more than a zombie. This difference, we could say, defines consciousness. It's that quality that gives us a sense of self, of what we are feeling, of introspection, examining and questioning our place in the world. It is, perhaps, what makes us amazed and entertained by magic tricks. It is what makes us laugh and cry. It is, you might say, what makes us human. Philosophers have longed to distil this essence of self-awareness for centuries. The exciting thing is that science is now, finally, giving us ways to probe the issue that involve more tricks than just thinking about it. And it appears that our scientific insights have killed the zombie. We can stand astride its corpse and declare that we will win in the end because we now see that consciousness must have a physical root and, consequently, will indeed succumb to science.

All of the work scientists have done on consciousness so far has led us to a handful of models that seek to exemplify what is going on inside our heads. Two are considered most promising.

One is the global neuronal workspace theory, a combination of insights from psychology and neuroscience. It suggests that all the inputs from the outside world – touch, taste, vision, hearing and so on – are first processed unconsciously. Very few of these inputs will get your attention; this only happens when there is enough subconscious processing going on to trip a switch that activates the areas of the brain concerned with conscious processing. Neuroscientist Daniel Bor describes it as 'a spotlight on a stage, or scribbles on a general-purpose cognitive white board'. Put simply, it's putting our short-term working memory to use – although those memories last only a couple of seconds, it's long enough to draw on them when necessary.

Its chief competitor is known as information integration theory. This model puts consciousness into the language and framing of information theory, creating datasets that add up to more than the sum of their parts. Its originator is Italian psychiatrist and sleep researcher Giulio Tononi. He is a controversial figure in many ways – though his theory is barely on its feet, he has declared that it could lead to a universal consciousness meter that would measure the 'level' of consciousness of anything from a worm to a computer network. However, information integration theory is about the whole network of neurons, and makes no attempt to explain what is going on in the individual physical structures of the brain. That means it doesn't much lend itself to the kinds of simple experiments used to test the global workspace theory. That said, it has some heavyweight fans. 'It's the only really promising fundamental theory of consciousness,' Christof Koch told *New York Times* writer Carl Zimmer.

In the end, though, we have to admit that decades of development have resulted in theories of consciousness that are still somehow unsatisfying. Psychologists and neuroscientists are, in many ways, like Darwin aboard the *Beagle*: they are still gathering specimens and making observations of interesting things done by the brain. They haven't yet, if we're being honest, got very far in pulling it all together into a coherent theory, a simple idea that explains the subjective experience of being conscious of what is around us, of thinking about things, of how the stuff of our brains creates a different experience from the zombie's existence without awareness. And that is exactly what led researchers to kill the zombie.

The leader of the zombie-hunters is unquestionably Tufts University philosopher Daniel Dennett. His strategy is remarkably simple. Perhaps there is, he suggests, no such thing as consciousness, that this ongoing awareness and sense of thinking about the world is actually an illusion. Perhaps our brains are fooling us into thinking there's some overarching narrative to our existence.

In 1991 Dennett published a book with an audacious title. *Consciousness Explained* was greeted with charges of hubris, but perhaps the detractors should have waited. In the book's 'Appendix for Scientists', Dennett made a prediction that, if his theory was right, we should be blind to many subtle changes in our environment. Change-blindness would exist, he said, because the conscious visual experience is not a true reflection of what is actually in front of individuals.

Dennett's idea is similar to the premise of the movie *The Matrix*, where humans have a conception of reality that is actually a carefully stitched simulation fed directly into their brains by a race of machines. In Dennett's view, there are no machines, only the brain. But, just as the machines' simulation sometimes has glitches, if we look carefully enough at our world, we'll see the brain's stitches. It turns out he was right.

Ronald Rensink has done a lot of the work to prove Dennett's hypothesis. He has carried out a series of experiments that show people missing seemingly obvious things right in front of their eyes. To understand why, we can start with the issue of foveal saccades.

The evolution of the eye and the visual processing system has had to cope with a number of efficiency measures, but perhaps the most remarkable one is that, even without taking blinking into account, for around four hours of every waking day there is no visual information being processed by your brain. That's because your retina takes in a full image of the world on a patch of densely packed photoreceptor cells that is about one millimetre in diameter. This is the fovea centralis, which records detail and colour from the world around you. The thing is, it only takes that reading from an area that's about the size of your thumbnail held at arm's length. Your vision captures everything else in front of you at that moment at a much lower resolution, and in monochrome. Go ten degrees off centre and you're capturing about 20 per cent of that maximum amount of visual information. In other words, most of what you see is recorded in a blurred black-and-white image.

The reason you're not aware of having such a 'lo-fi' view of the world is because your eye is constantly flitting about, capturing as much of the visual field as possible on the fovea's receptors. Roughly three times a second, for about 200 milliseconds each time, you record a high resolution image, and then your eye moves again. In between these saccades, or jerky movements, your brain turns off in order to prevent you registering the blurred image of the movement. In a paper published in *Trends in Cognitive Sciences*, David Melcher and Carol Colby showed that spending 100 milliseconds 'offline' roughly 150,000 times per day adds up to four hours of blindness. You don't notice it because your brain stitches together the processing it *has* done, creating the illusion of seamless visual perception. But that's nothing compared to the illusions exposed by change-blindness researchers.

Working with his colleague Daniel Simons, Dennett has per-formed some stunning (and hugely entertaining) experiments demonstrating how the smooth flow of our visual consciousness is an illusion. Starting with photographs that swapped between two views, for instance, only 50 per cent of subjects noticed when the heads of two men were swapped. No one noticed when the men swapped their differently coloured hats. Our 'sparse visual representations' meant that when subjects watched a film of an actor rising from a chair, and changes in camera angle were used to swap one actor for another, 67 per cent of people didn't see the change. The same thing happens in the real world. In one classic experiment, an actor stops someone on the street to ask for direc-tions. During their conversations, two other actors carrying a door rudely barge between them. While the door obscures the subject's view, a fourth actor replaces the first. In roughly 50 per cent of cases, the subject then carries on giving out directions, oblivious to the fact that the person they are now talking to is different from the first.

It's not because the actors appear to be similar: even if those two actors have different clothing and haircuts, and different heights, builds and voices, half the time, people just don't notice the change.

You can even exploit change-blindness. Film editors, like magicians, use tricks and distractions. Edward Dmytryk's seminal book *On Film Editing* makes it clear that sometimes you can make an audience blink, which gives you one-fifth of a second to change the camera viewpoint, or the focus of a scene, without any of the viewers noticing. He suggests the sound of a door slamming, but any sharp sound – a gunshot, for example – will do it. 'The cutter makes his cut as the viewer's eyes blink or are caught by the movement on the screen, much as a magician masks a move requiring camouflage by distracting the eyes of his audience with the broad sweep of his cape or a sharp movement of his "decoying" arm,' Dmytryk says.

Even movie editors are change-blind, though. That's where those cinematic howlers come from: there's a scene in the movie *Goodfellas*, for example, where a child is playing with a set of blocks. As the cuts come and go, so do the blocks – they change colour, or are stacked in different orders. In another scene from the same movie, a loaf of bread mysteriously disappears. Clearly no one noticed before the film was released, and most cinema-goers didn't notice, either. There's a scene in *The Wizard of Oz* where Dorothy's ruby-red slippers turn black for a moment. In the movie *Avatar*, one scene features golf balls that appear to move around the green of their own accord.

Such observations are unquestionably fascinating and fun, but they have a serious side. Skilled experimenters can use them to show that we are not paying proper attention to the world, we have little memory for the details of what is going on around us and we simply don't see what we think we see. Our conscious experience is not at all what we think. Consciousness has all the traits of something that evolved to give a 'just enough to survive' outlook on the world; it is a product of our senses, nothing more, nothing less. It is certainly not appearing to be something extra that is 'on or off' – human or zombie. It's more like a sliding scale. And that has significant implications – not least for the animals with whom we share the planet.

On 7 July 2012 a group of consciousness researchers gathered at Churchill College, Cambridge. They weren't philosophers, but cognitive neuroscientists, neuropharmacologists, neurophysiologists, neuroanatomists and computational neuroscientists. Together, they made a 'Declaration on Consciousness'.

The subject of their declaration was the abundance of new evidence concerning the 'neural correlates of consciousness'. The signals we can read from the brain, which tell us something about the subjective experience of the creature whose brain is being studied, are now showing that emotions and feelings abound in non-human animals and young humans. Invertebrates such as insects and octopuses have them. Birds have them: 'evidence of near human-like levels of consciousness has been most dramatically observed in African grey parrots,' the researchers declared. Zebra finches clearly experience REM sleep – they dream, in other words. Magpies can recognise themselves in the mirror just as well as great apes, dolphins and elephants do.

With all the evidence to hand, the researchers made their statement: 'non-human animals have the neuroanatomical, neurochemical, and neurophysiological substrates of conscious states along with the capacity to exhibit intentional behaviours… humans are not unique in possessing the neurological substrates that generate consciousness. Non-human animals, including all mammals and birds, and many other creatures, including octopuses, also possess these neurological substrates.'

They signed their written deposition, slightly oddly, in the presence of Stephen Hawking. But perhaps it was a good choice. Who would claim that Hawking, a brilliant cosmologist, isn't conscious? He is undoubtedly very much aware of his surroundings, capable of feelings such as joy and sorrow, and a cogent and fearsome thinker. Strip him of the technology that allows him to communicate, however, and of the human carers that meet his physical needs, and it might be possible to plausibly deny his consciousness.

That is why this area of research is so important. An understanding of consciousness is key to relating properly to animals, but it will also help us face our problems with the biggest human dilemma: death.

In 1985, when her husband was hospitalised with pneumonia, doctors asked Jane Hawking if she would like his life-support system turned off. Stephen Hawking was not yet a celebrity figure – *A Brief History of Time* was still an unpublished manuscript – and his diagnosis of motor neurone disease had robbed him of any significant life expectancy. He was in a drug-induced coma; the doctors were open to Jane ending it there.

Just occasionally, our medical skill, our ability to keep people alive, can be seen as a modern curse. Death already stalks us through life. We are among what is a very small number of animals who live with an awareness that we will one day die. It's made even more difficult when we are put in control of the timing. Perhaps that is why we find the coma patient – the seemingly dead living among us – so hard to handle. In the presence of a patient lost to a coma, unreachable and unable to reach out with any form of communication or gesticulation, we are reduced to our own peculiar paralysis. We don't know what is the 'right' thing to do.

Jane Hawking said no. Others haven't, though. Let's hope they haven't heard about Adrian Owen's recent discovery that supposedly disastrous brain damage, even when it results in an apparently 'permanent' vegetative state, can occasionally leave a patient in a conscious state.

Owen's team of researchers assessed a man who, while meeting all internationally agreed criteria for being in a vegetative state, was able to generate 200 responses to direct commands. He couldn't communicate with the outside world, but Owen's electroencephalogram (EEG) reading of the electrical activity in his brain made it clear that he was responding just as you or I would if we happened to be bound and gagged. 'He's probably as conscious as you or I,' is how Owen put it to *New Scientist* reporter Chelsea Whyte.

It's worth noting that this was just one patient in a study of nineteen; only three of those vegetative patients showed reliable signs of consciousness – and researchers continue to debate what constitutes a reliable response – and, of course, what constitutes a conscious response. Nonetheless, there is clearly something worth exploring here. Then there is the extraordinary discovery that coma patients show an emotional response to music that is played to them. Again, they can't show it outwardly, but their heart rate changes in exactly the way yours or mine would. We must be cautious about false inference – we don't know that the heartbeat change necessarily means these patients have any degree of consciousness – but it's another area ripe for exploration. If these people are really just 'locked in', that changes the way we must treat them. Owen suggests the first thing to find out is whether they are in pain. Then we must use technology to allow them to communicate. Brain–machine interfaces are becoming ever more adept at reading and interpreting brain activity; we might be only years away from holding a meaningful conversation with a coma patient. After all, even though he wasn't in a coma, no one should have been able to talk meaningfully to Carissa Philippi's patient, a man known as R.

R was a fifty-seven-year-old college-educated man whose brain had been severely damaged by an episode of herpes simplex encephalitis. The virus had destroyed his insular cortex, anterior cingulate cortex and medial prefrontal cortex. Those three areas of the brain are vital to self-awareness: according to the experts, R should have been a zombie.

He clearly wasn't. When the researchers asked him for his definition of consciousness, R replied with a cogent take on the issue: that it was the body's 'awareness and reaction to what's going on in the environment around it'. The researchers asked him, 'Do you think the sense of self is like a concept?' 'Yes,' he said, 'it's an idea in your brain.' Despite lacking the brain material thought essential for consciousness, R certainly didn't seem like a zombie.

R can't taste or smell, and he suffers from severe memory loss. However, he's not in bad shape, all things considered. The Iowa researchers carried out a series of tests and imaging studies on him. They found that his intelligence is in the normal range, and he recognises himself in the mirror. One of the researchers surreptitiously put black eye-shadow on R's nose, and he showed no hesitation in wiping it off when presented with his reflection fifteen minutes later (that was sufficient time for him to forget completely the researcher's contact with his face). He has a sense of agency over his actions: he knows what effects he causes. He can't, for instance, tickle himself; no one with a sense of their own agency can. When asked to do a little introspection, he reports he has a sense of himself. The researchers' conclusion is stark and straightforward: we have to stop thinking about consciousness as something that sits in a single 'higher', more evolved part of the brain, they say. 'R is a conscious, self-aware, and sentient human being,' the researchers declare; self-awareness, they conclude, 'is likely to emerge from distributed interactions among networks of brain regions.'

In some ways, that's not surprising. We know that the prefrontal parietal network, which seems to play a central role in conscious processing, contains the brain structures that have the most connections to other regions of the brain. Consciousness is, in that way, rather like the Internet. It, too, involves widely distributed processing, with no one place in particular tied to any one task. That is not to say there aren't particular tasks carried out at particular places, but, as R shows, the brain can adapt to injury – and so can the Internet. People have tried to turn it off by destroying parts of it – there was a huge attack in 2007, for instance, which aimed to overwhelm the thirteen servers that carry vital data. Two servers went down, but the other eleven carried on with business as usual. Having survived the attack, the system has been modified so that the servers also act as mirrors for each other: if one goes down, another takes over its role.

There are those who believe that the Internet could eventually exhibit a form of consciousness. If that seems a little far-fetched, we'll soon know for sure. We're about to start building artificial brains, and they stand an even higher chance of lighting up with some kind of conscious thought. Before we explore that frontier, though, we need to consider a rather provocative question.

'Can Machines Think?' It's a good question, which is why Hugh Loebner put it on his gold medal. Loebner is quite a character, by all accounts. He is a tireless champion of the rights of sex workers and, if you believe the press, a control freak obsessed with achieving Artificial Intelligence (AI). It's ironic, then, that the annual Loebner Prize is a competition that has serious AI researchers frothing at the mouth.

The premise is simple. Loebner will give $100,000 to the creator of a computer that can hold a typed conversation that is indistinguishable from conversation with a human. The prize comes with an 18-carat solid gold medal, on one side of which the face of Alan Turing appears, accompanied by Turing's provocative question.

The question is taken directly from a paper Turing published in 1950, 'Computing Machinery and Intelligence'. Turing outlined the idea that a sufficiently proficient machine, appropriately programmed, might be able to hold a conversation – via teleprinter – in a way that made it seem human. He called his suggestion the 'Imitation Game' and proposed it as a thought experiment (initially, at least) to address the question of whether the human brain does things a machine can't.

It is hard to know whether the Turing Test is of real value, but the annual jamboree surrounding Loebner's competition, which first ran in 1991, has certainly exposed the paucity of our achievements in AI. No one has come close to winning the gold medal, and the conversations conducted by each year's best entrant (winning a bronze medal and $2,000) are almost farcically odd. Loebner's insistence on carrying on with the competition, despite the unimpressive results, has caused AI pioneer Marvin Minsky to offer $100 to anyone who can persuade Loebner to put it out of its misery. Loebner, though, is having none of it. (He is a difficult man to deal with, by all accounts. An article in *Salon* from 2003 puts it beautifully: 'By tradition, three things happen at the conclusion of every Loebner contest: The winners take their prizes and run for the nearest exit, Hugh Loebner basks in glory, and the hosting organization takes a solemn oath: "Never again."')

The test wasn't Turing's first foray into machine intelligence. In 1948, while still based at the National Physical Laboratory in London, he wrote a fascinating paper he called 'Intelligent Machinery'. It describes an 'unorganised machine', where artificial neurons are linked by random connections that can be modified as necessary. He showed that, if the network was sufficiently large, it could perform all the functions of a general-purpose computer. Here, Turing suggested, was a possible link between the messy human brain and the high-level processing it manages to carry out.

It was not well received by his boss, Sir Charles Darwin (grandson of the famous Charles Darwin), who called it a 'schoolboy's essay' and left it to languish, unpublished, until fourteen years after Turing's death. The trouble was, computing machines were simply not advanced enough for anyone to take the possibility of mimicking human intelligence seriously.

In many ways the Turing Test misses the interesting point about machines that can think – perhaps deliberately. Biographer Andrew Hodges suggests that the Turing Test concept is a dodge that 'allows Turing to avoid any discussion of what consciousness is.' Perhaps to dabble in machine consciousness would have been a step too far for someone who was already seen as a dangerous maverick. After all, that 1948 paper does begin with what must have sounded like a ridiculous proposition.

'I propose to investigate the question as to whether it is possible for machinery to show intelligent behaviour,' Turing wrote. 'It is usually assumed without argument that it is not possible.' Over half a century later Pentti Haikonen answered Turing's question. He didn't bother with arguments about machines that display intelligent behaviour. He built one, and then he taught it the meaning of pain.

Pain is a curious phenomenon. It is, Haikonen points out, the result of a signal indicating some kind of damage to cells. In the brain a pain signal looks no different from a signal from the visual or auditory systems. And yet, while those signals can be processed without much attention being paid to their content, pain signals disrupt attention. They also modify our behaviour – they do anything to make us find a way to minimise the damage being inflicted. As he writes, 'This global disruption of attention is necessary as the pain signal itself does not know what should be done to stop the damage.' That's why it broadcasts the message to every system of the body and brain, disrupting whatever is going on and making us perform reactions – screaming, writhing, jumping away. 'I consider this disruptive broadcasting as a fundamental property of pain,' Haikonen says. And that's why he decided to build a robot that would react to pain.

When Haikonen beats Experimental Cognitive Robot, or XCR, it retreats. He can even teach it to be scared of things associated with harm. The astonishing thing is that it has no computer onboard. There are no microprocessors or programs; it's just electrical circuit components, such as wires, resistors and diodes – the same kinds of things that Turing built his first computers from. When Haikonen presents it with a green object and asks its colour, XCR responds with the correct answer. He then taps it on its back end with a pen. 'Me hurt,' it says in a plaintive voice before moving away from Haikonen's pen. Through its circuitry – not through any programming – XCR associates the colour with a negative emotion. 'Green   bad,' it says when a green object is subsequently placed in front of its eyes. And then it backs away.

There is pleasure in XCR's life too: Haikonen can give it desires. Stroking the robot on its top side produces a 'good' association with whatever is in its field of vision. The robot then moves towards it and uses its arms to embrace the object.

Haikonen believes his approach to consciousness will pave the way for sentient robots that have inner speech and mental imagery as well as emotions. They certainly won't be zombies; these robots, he says, will want to have fun.

Now that we are talking about want, desire, learning, hurt and pain, it certainly seems as though we have created a primitive consciousness, whatever the theorists might say. Somehow, the experimental approach to understanding consciousness – Haikonen's is but one of many such approaches – seems much more promising than what the theorists are achieving. Which makes it all the more exciting that we suddenly find ourselves in a position to build a whole brain.

A 2013 editorial in *Nature* announced the sea-change in outlook: 'technologies have developed far enough that it is now possible for us to imagine a day when we will understand the murky workings of our most complex organ: the brain.' Though that day remains distant, 'scientists are no longer considered crazy if they report a glimpse of it on the horizon.'

Over the next decade researchers from more than 130 different institutions will work together to build an unprecedented simulation of the human brain within normal digital – one might say organised – computers.

It's a daunting project. At first glance the human brain is a squishy lump of matter, but a close inspection will reveal what is perhaps the most complex object in the universe. It is composed of cells called neurons connected by gossamer-like threads composed of axons. Electrical and chemical signals travel between neurons; around a hundred billion neurons, each with somewhere in the region of 7,000 connections to other neurons, somehow come together to create consciousness.

It's possible to recreate this because three-dimensional maps of the brain created with electron microscopes can give nanometre-scale resolution. Neurons and their axons aren't big – neurons are only 20 micrometres in diameter – but that's plenty big enough when you can map things one thousand times smaller. We are gradually pulling together the information needed to create the neuronal map of a mouse brain – that's roughly 75 million neurons. The human brain's 86 billion neurons is a long way beyond that, but not so far that we can't set our sights on it.

The ambition is breathtaking. It will be possible, with advancing technology and the billion Euros of funding that has already been secured and started to flow into research labs in Europe, to simulate the human brain within ten years. Computing power doubles every eighteen months, and the massive computing power required for the Human Brain Project – roughly a thousand times what is available now – should be online by 2023. That will be enough to model the detail of neuronal connections, and the way ions flow in and out to pass signals between both single neurons and different areas of the brain.

The stated goal of the Human Brain Project is to investigate what happens when a brain goes wrong, in afflictions such as Alzheimer's disease and Parkinson's disease. The unspoken promise is that a silicon brain may well show a form of consciousness. Which brings us back to the zombie.

We used to think consciousness was all about the cortex, that this recent evolutionary addition to the brain was the root of consciousness. However, we now know that creatures without a cortex make conscious decisions and display emotional states. As Mark Bekoff and Jessica Pierce observed in their book, *Animal Justice*, 'We are not the only moral beings.' The Cambridge Declaration on Consciousness makes that crystal clear: 'The neural substrates of emotions do not appear to be confined to cortical structures… Wherever in the brain one evokes instinctual emotional behaviors in non-human animals, many of the ensuing behaviors are consistent with experienced feeling states, including those internal states that are rewarding and punishing. Deep brain stimulation of these systems in humans can also generate similar affective states.' Consciousness happens in the brain – all over the brain. Which means that our built brains might show it arising, killing that zombie dead.

As Stanford University philosopher Paul Skokowski demonstrates in his essay, *I, Zombie*, 'Microphysical and functional duplicates of us living in a duplicate world will have conscious

experience just like the beings that inhabit this world: *us*.' This
means, he points out, that zombies – duplicates without any
conscious experience whatsoever – are impossible. 'So please,'
Skokowski adds, 'do not bother weeping for your zombie duplicate; after all, he or she will be feeling the same pain as you.' It's
an elegant rephrasing of what Patricia Churchland calls, slightly
less elegantly, the 'hornswoggle problem'. Chalmers' zombie
argument , she says, is 'hornswoggle': bamboozling, sleight of
hand, no different to the tricks that Gustav Kuhn performs.
Why? Because we might well solve the easy problems of consciousness: how we form memories that we can dredge up into
our consciousness; how our visual perception becomes a conscious observation, a noted sighting; the difference between
unconscious sleeping states and the experience of being awake;
how we pay attention to something; how pain is sometimes only
experienced once we have had our attention drawn to what is
causing it – and then find there is no 'Hard Problem' left to
solve.

Churchland compares the objections people have had to this
notion with the objections people had to other revolutionary
ideas. 'When people were told that the earth moves they thought
this was hilarious; it was ludicrous; it was inconceivable,' she says.
Perhaps the best analogy is the problem people had seeing light as
an electromagnetic wave. Light had religious and emotional significance, and the idea that it came from the same phenomenon
that caused magnetism or static electricity was almost demean-
ing. The idea that our consciousness is a product of electrical
interactions between neurons is similarly problematic. But we'll
get used to it. In just a few decades we'll marvel at the problem
people had with consciousness, she reckons.

And the zombie will be dead. Interpretations of the psychology
associated with brains, coupled with the increased understanding

of what goes on at the cellular or even molecular level, will give us the insights equivalent to our modern understanding of the partnership between electromagnetic waves and the phenomenon we know as light. The neuronal firings and the consciousness are inseparable; the zombie, the collection of neurons without the consciousness, simply vanishes in a puff of logic. 'They're not two things embracing each other, they're actually just one thing, looked at from two different points of view,' is how Patricia's husband, neuroscientist and philosopher Paul Churchland, puts it.

It is time to let go of the rigid definitions and embrace the complexity. Our over-confidence in interpreting outward signs or simplistic criteria have led us down a blind alley with our investigations of consciousness. We became convinced that there must be some secret self, some hidden extra for true consciousness to exist. But in the end, all you seem to need is some kind of information processing equipment that works in a similar way to the jelly-like mass inside our skulls. If you've got what we'd call a brain – whatever form it takes – you'll exhibit some kind of consciousness. Not all of the brain needs to work for conscious activity. In some cases, relatively little of it needs to work. What's more, different-sized brains are likely to exhibit different kinds of consciousness, but we aren't the arbiters of the cut-off point where an organism isn't conscious. Blue light can't tell red light it isn't light: it's a spectrum, a spread. And to think, we are about to build ourselves a brain, perhaps creating a new consciousness somewhere on that spectrum, a consciousness the like of which has never been seen on Earth. Could there be a more exciting moment?

To move us on, here is an observation that neatly sums up the dilemma of neuroscientists wanting to understand what makes us the people we are: 'Currently, invasive procedures, such as decapitation, are required for most genomic analyses of brain processes, ruling out human studies.' Yes, it would be a boon to know what's going on in there, but decapitating people to find out is probably going too far. But is decapitating animals any better?

The quote comes from an article by Samuel D. Gosling and Pranjal H. Mehta in a book called *Animal Personalities*. Even the researchers working in this field aren't always comfortable with the term 'personality'. Some prefer to talk about animal temperament, behaviour types, coping styles or predispositions. Others choose the language of medicine, referring to an animal's 'behavioural syndromes'. What they all mean, though, is personalities.

As we are about to find out, individual animals have consistent and characteristic default responses to situations or ways of interacting with their environment and the other creatures within it. Most people accept personalities in their pets. Every dog-lover embraces the idea, and anyone who has owned more than one cat at least suspects that they have distinct personalities. Researchers looking across the animal kingdom have confirmed that our domestic pets are only the tip of the iceberg. We now know of vivacious donkeys, introverted octopuses, bubbly rats, sociable pigs, arrogant sticklebacks and timid funnel-web spiders. And once you know about that, it's very hard to decapitate anything. This field, still only a decade or so old, is sowing the seeds of its own destruction: discovering that animals have personalities as individuals and culture in their various groups makes it so much harder to do experiments on them. Much more than that, though, it is making us realise that human beings, for all our gifts, are nothing special.

2

THE CROWDED PINNACLE

Human beings are nothing special

Animals, whom we have made our slaves, we do not like to consider our equals.

Charles Darwin

You have almost certainly heard the seven-note riff of 'Seven Nation Army' by the White Stripes. Jack White, the band's singer and guitarist, composed it in 2001 when he was warming up his guitar before a concert at the Corner Hotel in Melbourne Australia. He thought the simple melody was 'interesting', and noted it down, later building a song around it. The really interesting thing is that, though nobody thought much of it at the time (a record company executive in the room said he actively didn't like it), it is now known across the world. And not because of the song's commercial success.

Washington-based writer Alan Siegel has charted the rise of the riff. Fans of the Belgian soccer team Club Brugge KV started singing it at a bar in Milan in October 2003. They continued singing it on their way to the stadium, and on their way out after the game. They sang it during their home games, and the club decided to adopt the tune as their own, broadcasting it over the PA at their stadium when the home team scored.

Italian visitors to Bruges took it back home, then Italian soccer stars took it up and led renditions from the stage during a Rolling Stones concert in Milan. Somehow, it then crossed the Atlantic and began to feature in huge American sports events such as NBA and NFL games. Back in Europe it was a ubiquitous feature during soccer's Euro 2008 competition. White was delighted. It had spread as far as the songs of the humpback whale.

In the same year that White composed his most famous riff, Luke Martell and Hal Whitehead published a paper titled 'Culture in Whales and Dolphins'. This was no dalliance with a cute idea: it is a monstrous seventy-four pages long with ten pages of references.

The pair considered three types of cultural spread. The first was 'rapid spread': a novel, complex behaviour spreading through a segment of the population. Then there was 'mother–offspring': cultural practices passed down the generations. Finally, they looked at the 'group-specific' culture: differences between groups that could not be explained by genetics, environment or the structure of their bonded groups.

An example of rapidly spreading culture comes from the songs of the male humpback whale. All the males sing the same song, which can be as short as five minutes long, or last for nearly half an hour. They are different every breeding season, but whales thousands of miles apart sing identical songs. For example, the songs on Maui in Hawaii and the Mexican Islands Islas Revillagigedo, more than 4,000 kilometres apart, evolve in exactly the same way from breeding season to breeding season. The only explanation is a cultural transmission; as a group hears the new variation, it takes it up – just as a group of football fans will take up singing a catchy tune.

Martell and Whitehead's 'mother–offspring' and 'group-specific' cultural traits are even harder to separate from human culture. Beluga and humpback whales follow their mothers on their initial journeys from breeding grounds to feeding grounds, then repeat the same journey for the rest of their lives. Some dolphins in Shark Bay, Australia, follow their mothers in the extraordinary practice of 'sponging', where they rip a marine sponge from the sea floor and fit it onto their beak like a glove. This allows them to rake through the sediment on the bay floor to root out the most nutritious fish without risking damage to their delicate beak. It's not a widespread practice – only five of the sixty dolphins in the Shark Bay area were observed to do it regularly. A few try but quickly abandon the practice. However, a female dolphin will become a sponger if her mother is one. And we all know of the human equivalents: girls who grow up to do things in exactly the way their mother did them. That might be baking a cheesecake from grandma's passed-down recipe, taking up a sport like gymnastics or enrolling in the same hobby group. With humans, it's certainly not just a female thing: male and female children have a tendency to go to the college their parents attended, or take up the family trade.

For group-specific traits, we could do worse than look at the killer whales in the seas around Vancouver Island. They exist in two distinct groups: the residents, who live in pods of around a dozen animals, and the transients, whose pods comprise just three whales on average. Each resident pod communicates in vocalisations that have distinct dialects that are passed down through the generations. The residents feed mainly on fish, while the transients are more interested in marine mammals such as harbour seals. The human equivalents are obvious. We have static and transient populations. We have cultural differences in preferred foods. And we certainly have dialects.

Though Martell and Whitehead broke through their peers' resistance to the idea of animal culture using whales and dolphins, plenty of other groups of animals show culture. Birds, like whales, have a musical culture, with songs learned from parents and neighbours through imitation (and sung in varying regional dialects). Over time, the songs slowly change, mirroring the evolution of human music.

Perhaps a classic example of animal culture is the extraordinary lengths to which the male bower bird goes to attract a mate. They build extravagant bowers to attract the females, weaving grass and twigs into a platform that they then augment with an archway made of leaves, studding the structure with brightly coloured berries, flowers and shells. This is not some genetic quirk; young males visit the bowers of older males early in the mating season, observing the intricacies of construction (and courtship). Then the young males go off and learn the trade, working together to build practice bowers of what is often rather crudely woven vegetation. On the odd occasion, the older males have been known to visit the younger generation during their first construction efforts and offer a helping beak. What's more, they don't have to be related, giving the lie to that evolutionary myth that animals only ever do anything to propagate their genes or provide themselves with food. We're used to seeing older humans regularly give up their time to coach a youth soccer team, help with a scout troop or teach musical skills. We don't worry that there's no apparent evolutionary payback. It's the same, as it turns out, with some animals. From play to grooming to attending funerals, animals value and enjoy the social bond as much as us. Many species show the same spread of personality traits that we do. It turns out that we're just not that different. Humans are nothing special.

Science has had trouble embracing this idea. Take Jane Goodall's experience, for example. 'When, in the early 1960s, I brazenly used such words as "childhood", "adolescence", "motivation", "excitement" and "mood" I was much criticised,' she wrote in a chapter of Peter Singer's *The Great Ape Project*. 'Even worse was my crime of suggesting that chimpanzees had "personalities". I was ascribing human characteristics to nonhuman animals and was thus guilty of that worst of ethological sins – anthropomorphism.'

For years – decades, even – Goodall, who started her research while employed as a secretary with no scientific training, was dismissed as unscientific, emotional and unreliable. A few people recognised the worth of what she was doing, though. In 1971, roughly a decade after Goodall started her research, David Hamburg of the Stanford University School of Medicine called it a 'once in a generation' endeavour 'that changes man's view of himself'. A couple of decades later, it had become clear that Goodall's work went deeper than that. The renowned palaeontologist and evolutionary theorist Stephen Jay Gould called it 'one of the great achievements of twentieth century scholarship' and 'one of the Western world's great scientific achievements'.

It seemed to make little difference to general acceptance of animals' place in the world. When Jeffrey Masson and Susan McCarthy published *When Elephants Weep*, a book about the emotional lives of animals, Masson shared his experiences of derision among academics for his research. An official at Sea World in San Diego said he would not permit the company to partner in Masson's work because it 'smacked of anthropomorphism'. Some people working with animals simply refused to talk to him about anything so 'unscientific' as animal emotion and feeling. But Masson, a trained observer of all animals – including humans – saw through the bluster. These people all worked with animals and their actions and activities belied their dogged calls for a detached, 'scientific' approach to animal studies, he noted. One researcher, who worked with dolphins on a daily basis, 'could hardly leave at night, so attached had he become to what he called his "subjects". Masson points out that it takes 'rigorous training and great efforts of the mind' to create such outward detachment from real feelings. The more extreme cases, he says, are suggestive of a psychiatric disorder.

Whatever the human pathology, there is a gaping hole in our understanding. 'As yet no prominent scientist has undertaken a sustained treatment of animal emotions,' Masson writes in the introduction to the book. That is just beginning to change.

Charles Darwin was one of the first scientists to stray into this territory. In *The Descent of Man* he wrote, 'It is often difficult to judge whether animals have any feeling for the sufferings of others of their kind … Mr. Blyth, as he informs me, saw Indian crows feeding two or three of their companions which were blind … I have myself seen a dog, who never passed a cat who lay sick in a basket, and was a great friend of his, without giving her a few licks with his tongue, the surest sign of kind feeling in a dog.'

Unfortunately that was 1871, and biological science was just starting to fall in love with hard facts and numbers. Such anecdotal reports were anathema to the growing number of biologists working to make their subject more like physics and chemistry. The 1883 publication of George Romanes's schmaltzy *Animal Intelligence*, an anthology of readers' tales and anecdotes didn't help. As Masson puts it, 'In reaction to this, science fled as far as possible in the opposite direction.' And so the subject was left to amateurs.

One of the first subsequent mentions of the dreaded word 'personality' came from a volunteer bird researcher based in Thomasville, Georgia. In 1922 L. R. Talbot described his experiences of putting bands on the legs of Georgian birds so that their habits could be logged and their habitats protected. The technique was simple enough. The birds would fly into traps for food. Talbot would then encourage the birds to fly through a small hole into a chamber where they could be handled for banding. Talbot wasn't very good at it at first, he admits, and a few birds lost feathers because of his clumsy handling. 'One Chipping Sparrow left his tail feathers in my hand when I tried to regain my hold after he had struggled and escaped,' he says. 'The Chipping Sparrow bore me no ill-will, however; he came back repeatedly and before I left Thomasville I had the satisfaction of seeing that his new tail feathers were almost full-grown.'

Tales of tolerant sparrows seem rather unscientific to our ears, but Talbot became very well known to these birds:

> It is not always easy to persuade people to do a thing of the desirability of which they have not become convinced; and sometimes a bird fails to understand why he is expected to go through a small opening which, apparently, leads nowhere and does not help matters in the least. But it is right here that different species, and even different individuals of the same species, show their 'personality', if I may use that term … The birds do not become frightened. Of course they become restless when they stop eating and first discover that they are imprisoned. They wander back and forth seeking an exit, but there is no reason to believe that they really have any fear. Birds that come back every day for two or three weeks, and sometimes four or five times a day, are not badly frightened.

Chipping Sparrow no. 22824 entered the traps six times in a single day. No. 22735 had a record of forty-three repeats, five in a single day. One Chipping Sparrow, no. 22849, repeated fifty-four times in twenty-two days. The birds would lie flat on their backs on Talbot's outstretched palm. Sometimes he had to force them to fly away. One blue jay used him as a plaything, lying on its back and then getting up, gripping Talbot's forefinger and letting itself fall until it was upside down.

Joyful blue jays are one thing. But fearless fruit flies? Yes indeed: even those 'lower animals' show distinct personalities.

Fruit flies can be categorised into rovers or sitters. Much like guests at a cocktail party, some will hang around the food while others gad about looking for something new and exciting. Though there is no such thing as a gene for a certain behaviour, there are genetic influences on personality. In the right – or wrong – circumstances, fruit flies can overcome their personality. Normally, a crowded situation favours the rovers, while sitters do well when there are fewer flies chasing the available food. If circumstances change – food becomes scarce, for instance – sitters will start to rove.

It's not that different from the way introverted people force themselves to mingle at a party. As any introvert knows, there are limits to what you can make yourself do. Not all animals are able to adjust their behaviour to suit circumstances. And that is why it is so important for even the animal-personality sceptics to embrace this emerging field. For starters, it has implications for medical research. Until we investigate the discovery that the health, lifespan and behaviour of laboratory rats, for instance, is linked to their personalities and the way individuals cope with strange conditions, our animal experiments may well be leading us up the garden path. We may, as we dig deeper, begin to question whether we should perform experiments on them at all.

'Animals from all taxa deserve ethical consideration,' is Jennifer Mather and David Logue's take on the situation. You can have hissing cockroaches square up to one another, they suggest, but no one is allowed to get hurt. Even that face-off might be too much: it's possible the roaches would be extremely stressed by the experience. Depending on their personality type, some will be as uncomfortable at the prospect of a brawl with a huge stranger as you would be. Cockroaches have feelings too.

Though science is uncomfortable with animals having personalities, it's an important consideration – conservation efforts have hitherto ignored the issue of personality in animals, for example, and that might have been a huge mistake.

At the turn of the century Susan Reichert and Ann Hedrick gathered themselves a couple of funnel-web spiders, one from the arid deserts of New Mexico and the other from damp woodlands of south-eastern Arizona. They also took a batch of egg cases from each habitat back to their lab and reared a new generation of spiderlings to adulthood, painting a different coloured dot on to each one's abdomen for identification. Then they looked to see what kind of personalities they could find among their spider subjects.

Press reports have called the funnel-web 'one of the world's most aggressive and poisonous spiders'. In some ways, that's justified. Most spiders scuttle away when disturbed, but not the funnel-webs: they 'may rear up and bare their fangs', according to reports. You certainly don't want to come into contact with those fangs: a bite will cause vomiting, breathlessness and convulsions. Unless an antidote is available, you might be dead in just two hours. But maybe we shouldn't tar them all with the same brush because not all members of the species are equally aggressive.

Reichert and Hedrick looked at boldness by allowing the spider to build itself a web. In the funnel-web this takes the form of a sheet with a tunnel at the back that leads down to a crack in the ground. This is where the spider can hide in case of attack. The spider can't just stay down there, however. An adult funnel-web spider needs to catch about 20 microgrammes of prey per day. Its web is not sticky, and so the creature has to be ready to pounce if an insect lands on it. That means sitting at the opening of the funnel and hoping you have enough time to retreat into the tunnel if a bird comes looking for a meal.

Survival, Reichert and Hedrick suggested, involves balancing aggression with appropriate caution, and the right balance will depend on geography. Aggression, for instance, is not only useful for meal-catching: in the New Mexico desert, prey is scarce, and there is serious competition for good places to make webs. To survive, you might have to fight other spiders for your site. What's more, there are few birds out to eat you in the desert, so caution is not a hugely important trait. The Arizona woodlands have the opposite scenario: plenty of prey and lots of hungry preda-tors. Reichert and Hedrick hypothesised that, because of this, New Mexico spiders would be bold and aggressive, and Arizona spiders would be more cautious.

They were right. When they used a camera-lens cleaner to blow a series of puffs of air on to a web, simulating the approach of a bird, the Arizona spiders drew back faster and stayed hidden longer than their New Mexico cousins. Since they had been raised from eggs in the laboratory, and had never encountered a bird in their sheltered lives, the researchers concluded that the tendency to caution or aggression had a genetic component.

The Arizona spiders were also slower to investigate a prey-like disturbance on their web. But not all Arizona spiders are alike. Some were even more cautious than others. Pitted against one another in a fight over a web (a fight that could last up to eighteen hours), the losers were always those who had been most timid in the face of a predator attack.

It's almost enough to make you feel sorry for a spider. The same thing has been found in other invertebrates, such as crickets and cuttlefish. In their amusingly titled essay 'The Bold and the Spineless', Mather and Logue catalogue a spectrum of invertebrate personalities that we ought to take in. For every sedentary fruit fly there is a bold, inquisitive damselfly. For every risk-averse field cricket there is an aggressive cockroach. Some squids are eager; others are consistently hard to engage.

It's unquestionably curious and fun, but there is a serious side to this discovery. A 2008 study by J. G. A. Martin and D. Réale in Gault Nature Reserve in Quebec, Canada, found that shy chipmunks would keep to areas of the reserve where humans are less frequently seen. The researchers pointed out that assessments of the impact of human presence on a population of animals have to take issues of animal personality into account. Faced with a pressured environment, the most aggressive animals gain the web, the nest or the hidey-hole. Usually they gain the reproductive rights. That's a problem because they make the worst parents, making the least effort to protect and feed their offspring. Having nothing but aggressive types in the locality puts a population's survival at risk.

It's not just about bad parenting. Personality traits such as aggression are linked to a particular set of genes, so outside pressures can reduce the variety of genes available to a population. That creates reproductive problems and can also create a dangerous uniformity in behaviour. If everyone is taking risks in foraging and other behaviours, there is a heightened chance that too many will be taken by predators and a population will collapse. Other scenarios favour the shy individuals, but this is equally problematic. Get too many of these in a population and you'll find no one is exploring new areas for food. At the first crisis – a dry spell, for instance – the population starves. Then there's the question of how they respond to chemicals in their environment: pollutants in rivers, for example. A 2004 study of the three-spined stickleback found that the chemical ethinyl estradiol, which is contained in birth-control pills and has been found in significant concentrations in waterways around the world, makes females exhibit more risky behaviour. Though that might seem like a good thing, it wasn't. They were taken by predators in much greater numbers than those in unpolluted waters. Personality, natural or acquired, can be a life or death issue.

It's worth noting that some animals deliberately seek out personality-transforming chemicals. In 1998 R. J. Cowie and S. A. Hinsley watched great tits feeding their newborn chicks for thirteen days. On days three to nine the parents brought their chicks a huge number of spiders. Subsequent studies have confirmed this behaviour: the deliberate choice of spiders when the chicks are around five days old. It doesn't matter what time of year, what kind of habitat the tits are living in, or whether spiders are abundant or scarce: parent birds will find a source of spiders and, for just a few days, allow their offspring to gorge on them. Why? Because spiders contain fear-busting chemicals called taurines.

In mammals, taurine, an amino acid, seems to be vital for proper brain development; taurine-deprived cats have problems with their vision, and human children who lack taurine are known to have a depressed IQ and impaired motor functions. Since the bodies of babies in or just out of the womb can't manufacture it, evolution has arranged for mammals to get some through their mothers' placenta and milk.

If you're not a mammal, taurine can be quite hard to come by. It would certainly be easier for the birds to catch caterpillars for their young: nutritionally speaking, they are broadly similar to spiders. The only significant difference is that spiders contain between forty and a hundred times more taurine. The reason behind a birds' instinct to gather taurine is unknowable, but subsequent studies have shown that ingesting it inhibits anxiety responses, making you bolder. 'Birds that had received taurine supplementation as nestlings, mimicking a neonatal diet high in spiders, were more likely to be risk takers,' according to Kathryn Arnold and her co-workers at Glasgow University. They also had better general and spatial learning abilities than the birds that didn't get the spider supplement. If none of the parents are of the type to go get spiders for their young, population collapse due to lack of risk-taking is a very real possibility.

We can't leave this subject without exploring animals' role in laboratory research a little further. The ethics of using animals at all is clearly on the table when we think of them as being bold or timid, disengaged or vivacious. But even if we do continue to use them in experiments, we have to ask whether we are using the ones with the right personalities.

In the early 1970s cardiologists Meyer Friedman and Ray Rosenman noticed that some of the chairs in their waiting room were badly in need of upholstering. The upholsterer who came to assess the job was puzzled by the pattern of wear: the chairs had worn out at the front of the seat and on the arm rests. He pointed this out to Friedman and Rosenman, and told them that chairs normally wear on the back rest. The doctors started to observe who was using the chairs and came to the conclusion that it was their cardiology patients – they were sitting, quite literally, on the edge of their seats. Metaphorically speaking it was also true: they were the most impatient of patients, never relaxing into the chair and frequently jumping up to see if their consultation was next. That impatience, drive and frustration, the doctors concluded, was the cause of the wear in the waiting room chairs, and was quite possibly linked to the patients' heart problems.

Friedman and Rosenman published a book in 1974: *Type A Behaviour and Your Heart*. Type A personalities are frantic, driven, hostile, always in a rush, often doing two things at once, never relaxed behind the wheel of a car but always trying to get to their destination quicker. They are aggressive and easily provoked into fights. That external drive and hostility is a result of an internal over-activity: their adrenal systems are pumping out far more hormones than normal. The result of this is scarred cardiac arteries and damaged cardiovascular systems. There has been much debate about Friedman and Rosenman's conclusions since they published their book, but it seems there is a strong link between Type A personalities and heart disease. So, if you're going to try to help these people with drugs tested on laboratory rodents, you'd better make sure the rodents are Type A too. Otherwise you'll be testing your drug in a biochemical soup that bears little relation to the human patient waiting (impatiently) for a fix.

The commercial breeding programs that have sharpened genetic traits in laboratory rats and mice mean there are rodents whose personalities are more or less suited to cardiac drug testing. A good bet are the 'Short Attack Latency'(SAL) mice. They are generally hostile and have been identified as displaying something similar to Type A behaviour. A bad bet would be the neophobic Sprague Dawley rats who hate being put in new situations. These rodents are much more suited to research into allergies and anxiety disorders, which are medically associated with shyness and inhibition in humans. Research on the physiological links to these behaviours is still in its infancy, but it makes sense to develop medicines that prevent or treat cardiovascular disease in the animals that are most like the typical human sufferer. As Sonia Cavigelli and her colleagues put it, 'if the goal is to understand the development, physiological substrate, and health implications of specific personality traits, then the most promising method will be to identify personality traits in rodents that most closely mimic human personality traits.'

If we are coming to terms with animal personalities – though we might prefer to call them behavioural syndromes – we also have started to embrace animal culture. The cultural life of whales is now more accepted than it was when Martell and Whitehead documented it so extensively, but it doesn't end there. Biologists have now seen cultural activity in animals as diverse as guppies and elephants. From the ingrained habits of the hunting humpback whale to the stay-abed indolence of the lazy meerkat, there are learned behaviours specific to each group. Some chimps will greet each other with a knuckle bump, others with a slap. Baboons may not go to the ballet, but as they grow up within their troupe they will learn a few lessons about the right way to do things. And that will almost certainly be different from what is considered right in other troupes. You'd soon know if an outsider joined the group. Which may be precisely the point.

Culture is a difficult term to pin down, partly because we all belong to a multitude of cultures. Some might be racial, others national or tribal. Then there's a commitment to a certain sport – football culture, for instance, and even the cultural life that goes along with supporting a particular team. There is religious culture that might define how you spend a Friday night or a Sunday morning. You might belong to a whisky-tasting club, another mark of a culture. Then there's musical taste: you might belong to a culture that appreciates jazz music or opera. You might belong to a drug culture that enjoys the effects of mind-altering substances.

The Oxford English Dictionary defines culture primarily as 'the arts and other manifestations of human intellectual achievement regarded collectively' and 'the ideas, customs, and social behaviour of a particular people or society'. Wherever you look for a definition of culture, one thing rings out. It is a pattern of behaviours that are shown by one group and mark them out from other groups. By that definition, we can have no hesitation talking about animal culture.

Animal culture first came out into the open in 1999. That was when a team of primate researchers – Jane Goodall was among them – cried 'enough' at the snobbish objections to the idea that animals have shared traditions. Between them, they had accumulated more than a century and a half of observations of chimpanzees, during which time they had recorded thirty-nine different behaviour patterns, ranging from courtship routines to grooming to use of tools.

The geographical distribution of these behaviours was striking. Chimps from the Taï forest of the Ivory Coast, for example, had a culture of using sticks to probe bee nests for honey. Chimps at the other five African sites under investigation didn't. Guinea's Bossou chimps were conspicuous in having no culture of rain-dancing. Those at the Gombe reserve in Tanzania were the only ones to customarily dip a stick into an ant's nest, then wipe the ants on to their hand before eating them. Taï forest chimps put the stick straight into their mouths. Gombe chimps were also the only ones to use a leaf to squash parasites during grooming. Taï forest chimps place the offending bug on their arm and smack it hard.

Then there's the issue of death. Death is responsible for much of the cultural attitude we carry with us through life. Plenty of the stories we tell, through books and plays and songs, are rooted in the human experience of death, and our reaction to it is perhaps one of the great definers and demarcators of culture.

Many animals seem to carry similar, if less developed, cultural baggage, when recognising the passing of those close to them. Here, of course, we have to rely on anecdotes and careful observation: as John Archer noted in 1999, in his book *The Nature of Grief*, laboratory experiments that deliberately induce grief in animals for the sake of research would be 'ethically unacceptable or at least questionable'. But for all Archer's study of human grief and its complexities, he is in no doubt that animals grieve over separation from infants or close companions. And, despite his misgivings, some experiments have been carried out and confirmed his hypothesis.

Teresa Iglesias, for example, performed the tests on western scrub jays: when she and her team put a dead jay into the back yard of someone's home, it was soon spotted. The jay that saw it would utter a call, and all other jays in the vicinity would stop their usual foraging behaviour to come and take a look.

When western scrub jays encounter a dead bird, they call out to one another and stop foraging. Iglesias watched a group of jays assemble around the body and begin a cacophony of calling. Only a day later did normal behaviour resume: clearly, the jays think the death of one of their kind is something that should give them pause. Taking things too far for many researchers' taste, Iglesias published the observations under the controversial title 'Western Scrub Jay Funerals'.

While we're on the subject of funerals, there is a remarkable tale in the book *How Animals Grieve* by anthropologist Barbara J. King. King relates the experience of primatologists Christophe Boesch and Hedwige Boesch-Achermann at Taï in Côte D'Ivoire when the chimpanzee they called Tina was killed by a leopard:

The Boesches found a dozen chimpanzees, six females and six males, sitting in silence around the body. Over the next hours, some of the highly aroused males performed displays around the corpse. Some touched Tina. In a period of eighty minutes, males Ulysse, Macho, and Brutus groomed the body for nearly an hour. Ulysse and Macho hadn't been seen to groom Tina at all when she was living; other males in the community had groomed her only for brief periods.

Eventually, Brutus, the most intelligent of the group, chased off the younger chimpanzees, who had begun to lark around near the body. However, Tina's younger brother Tarzan was allowed to approach the body, sniff and inspect it. 'Brutus recognised that Tarzan, alone of all the young chimpanzees at Taï, needed time to inspect his sister's body and to mourn over it,' King says. 'Tarzan mourned as part of a social community, because the alpha male of that community recognised his relationship with this sister.' King describes the ritual as a kind of wake. It lasted six and a quarter hours.

There are many odd stories of animals dealing with death in King's book. Some of them are hard to believe – like the tale of the hens that sought a human rescuer when one of their number fell into a swimming pool. Then there's the stories of animal suicide, one of a bear and another, reported in an 1847 edition of *Scientific American*, of a grieving gazelle. King is unsure what to make of either.

Other anecdotes seem more plausible, though. King tells the story of the mother of a stillborn dolphin, who attended to its body for days, pushing it through the water and defending the corpse from ravenous gulls while risking collapse by failing to feed herself. From time to time other dolphins would step in and swim around with the decaying carcass, keeping it moving, as if to show empathy for the grieving mother.

Not all chickens, goats, chimps or dolphins grieve in all

circumstances, King observes, just as humans are capable of experiencing the death of a loved one without any displays of emotion. 'The great lesson of twentieth-century animal-behaviour research was that there was no one way to be chimpanzee or goat or chicken, just as there is no one way to be human,' she says. Having said that, many animals do seem keen to pass on their ways of doing things.

One of the criticisms thrown at researchers who make claims for animal cultures involves the issue of teaching. The distinction between different chimp communities showed that these cultural norms were learned as the chimps grew up, just as British children learn to peel prawns while Singaporean children learn to eat them shell-on. But copying a behaviour is not the same as teaching one: for many researchers, only when animals are observed teaching others to mimic their actions – indicating an understanding that the action has a purpose – will they be satisfied that a culture is in place.

There have been various claims that animals have been seen teaching members of their community. The practice usually has to pass three tests to be accepted as teaching. First, the teacher modifies their behaviour when in the presence of an unskilled or naive individual. Second, the teacher derives no immediate benefit from its actions – it might even incur a cost. Third, the student acquires knowledge or skills faster than it would otherwise have done. With those safeguards in place, we know cultural knowledge and skills are not being passed on by accident through simple copying behaviours. And a few animals have passed the test.

We've already mentioned the senior bower birds who give bower-building tips to the younger males in their community.

Some ants, it turns out, are teachers too – they help others to learn how to find a new nest site and can even evaluate whether their students are good learners. Then there's the first birds to be seen teaching: the pied babblers, which use a peculiar call to train their fledglings in recognising and finding food sites. Most interesting of all are the meerkats who teach their young to handle scorpions.

Cambridge University researcher Alex Thornton discovered this behaviour in the South African Kalahari. Meerkats hunt alone, so there is no possibility of pups tagging along to learn how it's done. Instead, the adults will bring back items of food. When the pups start out, the adults present them with dead prey. But as they get more efficient at dealing with that, they change the strategy. With scorpions, which are difficult to handle safely, the adults will remove the sting but hand it over alive – the pups then have to learn how to kill and eat this prey. When they are good at that, the adults stop removing the sting, and let the growing pups negotiate the danger for themselves. It's probably a little tiresome for the adults: they have to hang around while the pups wrestle with the prey, and chase it down again when they lose it. In the end, though, their patient teachers will be rewarded with a large, competent group of hunters.

All these phenomena – personality, culture, teaching – point towards one inescapable conclusion: we're somewhere on the spectrum of the capabilities of life on this planet. We're not separate. It's fair to say – and somewhat obvious – that human culture is vastly richer than that of any animals, and shapes the Earth in ways that we are only just beginning to realise. But there's a reason for that, and it may have arisen by accident.

The biggest distinction between us and the rest of the natural world is the step change between animal communication and human language. We don't know exactly why this gulf exists. As Steven Pinker said in *The Language Instinct*, 'The first steps toward language are a mystery.'

It hasn't helped that this too has been a banned subject in the past. The Linguistic Society of Paris forbade the discussion of language evolution in 1866: Article 2 of its 1866 statutes said (it seems appropriate to show the original wording): '*La Société n'admet aucune communication concernant, soit l'origine du langage, soit la création d'une langue universelle.*' 'The Society will not admit any discussion concerning either the origin of language or the creation of a universal language.' *Vive la différence*, in other words: human language is both diverse and unique.

Though the Parisian ban was not enforced globally in any way, all scientists who research language seem to agree that, because prominent linguists chose not to engage with biologists and anthropologists, and were staunchly opposed to the idea that there could be any bridge between animal and human communications, it remained hugely effective for more than a century.

The divide was reinforced by Friederich Max Müller, the Oxford professor of linguistics who railed against Charles Darwin's newly minted theory of natural selection. Darwin had published in 1859; two years later Müller delivered a series of 'Lectures on the Science of Language' at the Royal Institution of Great Britain in which he derided the idea that anything derived from animal communications could account for the origins of human language. Language, he said, 'is the Rubicon which divides man from beast, and no animal will ever cross it ... the science of language will yet enable us to withstand the extreme theories of the Darwinians, and to draw a hard and fast line between man and brute.'

There are physiological reasons for that Rubicon. Our larynxes are low in our throats, and our pharynxes are long. The position of the larynx makes us the only animal not able to breathe and swallow simultaneously. No one agrees on whether this is an evolutionary advantage or disadvantage. It is simply a fact – and one that enables us to make sounds no other animal can. But it is not just about how air can move through our throats to allow us a unique range of vocalisations. It is also about our brains and, more interestingly, our minds. What gives us the ability to represent abstract ideas in words, then to play with them, stack them, throw them around at each other in complex ways that are almost always successful communications?

This new – in evolutionary terms – power is why Martin Nowak claims that 'language is the most interesting thing to evolve in the last several hundred million years.' It is, he says, something that changed the rules of evolution itself. Where information was only transferred by genetics, thanks to language we can accumulate and store knowledge, pass it on and accelerate the process of changing ourselves. Language led to the intellectual pursuits, which has led to our ability to augment our abilities with computers, to prevent and heal diseases that would have wiped most of us out by now, and to so drastically alter the planet on which our species plays out its existence. If we are unique, it is because of the accident of language.

Linguistic abilities notwithstanding, we see ourselves as elevated far above the rest of the animals at our peril. Our cultural transmissions have, after all, led us into many delusions and dangerous divisions. The biologist E. O. Wilson put our dilemma neatly. 'Contrary to general opinion, demons and gods do not vie for our allegiance. We are self-made, independent, alone and fragile. Self-understanding is what counts for long-term survival, both for individuals and for the species.'

Part of that process of self-understanding involves the

humbling acceptance that we are part of the ecosystem, inter-dependent with the other species, and not qualitatively different from them. And with that, we will move on to the controversial practice of mixing human and animal material. Are you happy with the creation of chimeras?

3

THE CHIMERA ERA

We are ready to make a whole new kind of creature

If your parrot says 'Who's a pretty boy, then?' that's ok. If your monkey says it, that's a very different matter
Christopher Shaw

I f you want to retain your admiration for the luminaries of the Enlightenment, you might want to look away now.

In 1656 Christopher Wren, architect, astronomer, surgeon and member of the Oxford University Experimental Philosophy Club, wanted to find 'a way to convey any liquid Poison into the Mass of Blood'. Wren started with opium, injected into a dog. It worked a treat, starting a litany of experiments in which everything from milk to dye was introduced into the veins of some hapless mutt, just to see what would happen.

By the end of 1660 the dog-injection experiments had a new home base. Charles II had chartered the Experimental Philosophy Club as the Royal Society, and in February 1666 Robert Hooke, the society's curator of experiments, led a team that performed the first dog-to-dog blood transfer. The team used a syringe, but

had bigger plans. Soon they were trying direct vein-to-vein transfusions. These didn't work, because of clotting, but artery-to-vein
transfusions did. In fact they worked too well: in the first experiments, they managed to transfuse so much blood (using a quill as
a canula) that the donor died. Eventually they managed a transfusion where the donor survived, but in that case the recipient died.
It was a steep learning curve.

Not one to be put off, another member of the team, Robert
Boyle, suggested a list of follow-up experiments. Recipient
animals, assuming they survived, were to be monitored to see
if the transfusion altered their behaviour, habits or any physical
traits. In November of the same year, with London still recover-
ing from the Great Fire, the experiments were performed before
an audience at the Royal Society. The observers included Samuel
Pepys, who went on to the Pope's Head tavern afterwards and
discussed the experiments with a Dr Croone. They came up with
some innovative ideas for future projects, such as 'the blood of a
Quaker to be let into an Archbishop'. There was much merriment,
but all good humour vanished when news broke that French scientists had stolen English thunder. The French, it transpired, had
succeeded in transfusing animal blood into a human.

The English scientific establishment was appalled at this
affront – appalled enough to alter the scientific record. We know
this because there are two sets of pages 489 to 504 for that volume
of the 1667 *Philosophical Transactions of the Royal Society*.

This confusing state of affairs arose because Henry Oldenburg,
the editor of the *Philosophical Transactions*, had been charged
with treason and was locked in the Tower of London when the
French manuscript arrived at the Royal Society. The publisher, a
Mr Martyn, was somehow entirely unaware of the English experiments, and published the French work without delay. When Oldenburg was acquitted and released, he was outraged by Martyn's

ignorance. He rounded up all original copies of that issue of the *Philosophical Transactions* he could find and destroyed them. Then he set about abbreviating the French work, and supplementing it with an editorial offering a detailed account of the English transfusion work. The tone was of a man who could never commit treason against his beloved nation: 'how long soever that Experimente may have been conceived in other parts ... it is notorious, that it had its birth first of all in England.'

Oldenburg did not succeed in destroying all original copies of the French work, however – which is why we have such a good account of the experiment of the animal–human transfusions of Jean-Baptiste Denys, physician to King Louis XIV, the Sun King.

Perhaps witnessing the excesses of a royal court at close quarters informed Denys's motivation for treating the sick with the blood of animals: it would be, he suggested, a purer substance than the blood contaminated by man's 'debauchery and irregularities in eating and drinking'. His first patient was a sixteen-year-old who had been tormented with a 'violent fever' for two months. Physicians had bled the boy more than twenty times, to no avail. The transfusion of 9 ounces of lamb's blood didn't exactly have him leaping out of his sick bed, but neither did it kill him.

The next recipient of lamb's blood was a forty-five-year-old sheep shearer, who was in rude health. So rude, in fact, that after receiving the transfusion he cut the lamb's throat, fleeced it, then went to the pub. Almost as soon as the reports were published, the British responded by putting 32 ounces of sheep's blood into Arthur Coga, a minister. Coga, whom Pepys described as a 'poor debauched man' and 'a little cracked in the head', received 20 shillings for his troubles. A month later, perhaps short of cash, Coga took another dose of sheep's blood.

Jolting you back to the twenty-first century, I doubt you would be comfortable about the idea of receiving an injection of sheep's blood – even for a lot of money. Mixing human and animal tissues is difficult for us to countenance. We don't even seem to like human-to-human mixing: only 4 per cent of the eligible UK population donate their blood, for instance (it's 5 per cent in the US). Some don't do it because it is inconvenient. Some hate needles. Some fear infection. Some find it distasteful to come into such close contact with the fluid that – all things being equal – should stay on the inside. Most, though, just don't even entertain the thought.

Organ donation is an even bigger problem. In the US the waiting list is 118,667 people long as I write. In the UK around 7,000 people are waiting for organ transplants, and one in seven of them will die because there are no suitable donors. That is why Sally Slater, a twenty-year-old with a sixty-three-year-old heart, has lent her voice to the campaign for people to have to opt out of the organ donation scheme, rather than opt in.

Slater's is a powerful voice, because her plight once captured a nation. In February 2000, at the tender age of six, she caught what her parents thought was a cold. As it turned out, the virus was much more deadly than that: it attacked the muscles of her heart. Doctors attached her to a machine that could help the muscles continue to pump blood around her body, but it was clear that this was not a solution that would work for long. At the end of March her father made an impassioned public plea for a heart donor. Sally was lucky. Three days later a donor was found, a middle-aged woman, killed in a road accident. The woman's family had been touched by Sally's father's plea, and the new heart was duly installed at the Freeman hospital in Newcastle-upon-Tyne. A week after that she woke up from the operation with a new chance of life. On 12 May her delighted face beamed from the front page of *The Times* newspaper. She was 'delighted' to be going home and was looking forward to eating her favourite food: her mother's carrot cake. Her father's campaign had prompted a surge in people registering to become organ donors and, fourteen years later, Sally is still healthy and still campaigning for others to be given the chance that she had. But what if one day soon that might not be necessary?

The day after Sally left hospital, two biomedical researchers filed a patent application. 'The invention', Thomas Ryan and Tim Townes stated, 'provides animals that produce cells, tissues, and organs of another organism.' Ryan and Townes had spotted that knocking out certain genes and inserting genetic material from an afflicted patient would mean you could rear a pig (or monkey or cow) whose heart, liver, pancreas, blood or skin cells were human, genetically matched to the recipient, and in every way perfect for transplantation. Before we explore how that is becoming reality – and it is – let's pause and explore the issue of whether mixing animals with humans is something we really want to do.

'The State is called upon to produce creatures made in the likeness of the Lord and not create monsters that are a mixture of man and ape.' You might think this quote comes from an angry bishop or a science-loathing evangelical preacher. In fact, it comes from *Mein Kampf*, Hitler's manifesto for national socialism. The revulsion at the idea of mixing humans and primates has been shared by many. But not, it seems, certain twentieth-century men of science.

The first scientist to suggest explicitly that primates and humans might successfully interbreed was Ernst Haeckel. At the turn of the twentieth century, Haeckel, one of the world's foremost experts on Charles Darwin's new theory of evolution, cited studies of animal and human blood as supporting evidence. He would no doubt have been aware of the seventeenth-century cross-transfusions between humans and animals. The practice had fallen into disrepute for a time, but transfusions were back in business by the time Haeckel was approached by amateur zoologist Marie Bernelot Moens. Moens, a teacher from Maastricht in Holland, wanted to prove evolution's truth by demonstrating the proximity as species of man and chimpanzee. If a female chimpanzee could be impregnated using human sperm, it would show there was no real barrier between the species.

Haeckel gave Moens his blessing and advised him to use sperm from an African man, as he believed these were the most primitive human specimens and thus most likely to successfully interbreed with the chimps. Moens began to put together an expedition but was quickly thwarted by public outrage. He was removed from his teaching post, his career left in tatters by his chimeric ambition. Perhaps that's why the venture that did eventually perform Moens's experiment was kept so quiet for so long. It was only in 2002, after a Russian historian of science had spent a decade sifting through the evidence, that we found out exactly what Ilya Ivanovich Ivanov had done.

Ivanov was artificial inseminator to the animals of the court of Imperial Russia. While the rest of the world dismissed the technique as unnatural, Ivanov developed tools and techniques that allowed him to become the leading insemination expert of his time. The imperial stud farm was in rude health, and Russia's farm animals went from strength to strength. By the end of the first decade of the twentieth century, he had successfully hybridised many species, including a zebra and horse. It was in 1910, at the International Zoology Conference in Graz, Austria, that he first mentioned crossing apes with humans.

Ivanov didn't act straight away. The rumblings that eventually led to the Russian revolution were making his situation as the Tsar's employee a little difficult. Then the First World War came and went. But in 1924 Ivanov found himself doing a little work for the Pasteur Institute in Paris and had the ear of Albert Calmette, one of the directors. Calmette provided the 'C' in BCG (Bacillus Calmette–Guérin), the new and highly successful vaccination against tuberculosis. He was keen to do more tests of his medical innovations on chimps and clearly thought that Ivanov's expertise on artificial insemination would be useful at a research station where chimps had consistently failed to breed. So he gave permission for Ivanov to travel to the Institute's primate research station in the province now known as Guinea-Conakry. Here, Calmette said, Ivanov should feel free to try his hand at inseminating chimpanzees with human sperm.

Ivanov had presented his proposal as a chance to 'provide extraordinarily interesting evidence for a better understanding of the problem of the origin of man'. The Russian Academy of Sciences agreed that it had 'great scientific significance' and 'deserved full attention and support', and gave him enough money to get to Conakry.

Clearly not everyone in Russia felt the same, though; one scientific well-wisher told Ivanov, 'Don't pay attention to all the gossip and rumours about your trip. To the devil with them!' The American Association for the Advancement of Atheism was also supportive: it promised to gift the experiment with $100,000 if Ivanov were able to produce a human–ape hybrid. The Atheists, it should be noted, were aware of the incendiary nature of the experiment and advised him to do things very quietly, keeping his head down. Ivanov's suggestion that he come to America to give a series of lectures was dismissed as premature: 'the best time to have you come to America to lecture would be after the first little anthropoid human shall have been born and ready for exhibition.' Word did get out, however. Ivanov received letters of indignant abuse from members of the Ku Klux Klan, among others.

Kirill Rossiianov's account of what happened in Guinea-Conakry makes disturbing reading. There was good reason why no successful breeding had occurred. First, the chimps were almost all pre-pubescent. Researchers had bought them from hunters and trappers who killed the adult chimps and sold on their young. Second, the primate research station was a sorry place, where chimps rarely stayed alive for long. In the three years the research station had been open more than 700 chimps had been purchased, but fewer than half lasted long enough to be transferred to the biomedical facilities in Paris.

Ivanov first arrived in February 1926 and stayed a month, but achieved nothing. He returned, with his son as an assistant, the following November. Eventually they found three mature female chimps. Ivanov carried out the insemination under the pretence that it was a medical procedure, in order to avoid the suspicion of Guinean staff to whom any form of liaison between human and chimp was an abomination. However, all his expertise in artificial insemination was wasted in these conditions. The procedure was rushed and slipshod. The sperm was 'not completely fresh' (obtained from a local man, apparently) and was injected only into the chimps' vaginas, rather than their uteruses, as Ivanov would have preferred. The chimps were held in nets that were twisted about their bodies to hold them still. Ivanov and his son each had a Browning in their pocket in case the chimps got free. It was, Rossiianov notes, 'brutal and hurried ... like it was a rape'.

And it was unsuccessful. So was their second attempt, on a chimp called Black who was put to sleep with chloroethyl.

Ivanov's follow-up plan was to inseminate Guinean women with chimp sperm – without their knowledge or consent. Astonishingly, he did get permission from the governor. But a week later the governor changed his mind: 'a terrible blow', as Ivanov records it in his diary. Clearly, this gave Ivanov no pause; on his return to the Soviet Union, he arranged a programme in which Soviet women would be inseminated with sperm from the country's one captive orang-utan. At least one female volunteer came forward, but the animal, called Tarzan, died before any experiment could commence.

That was the end of Ivanov's attempts to create a human–ape chimera. While the researchers waited for the delivery of five chimpanzees, he was publicly accused of political subversion. As a former employee of the imperial court, he was always in a precarious position, and on 13 December 1930 Ivanov was arrested, and then exiled for five years in what is now Kazakhstan. One day before his restoration, which had been mandated by Joseph Stalin, Ivanov suffered a fatal stroke.

There are good reasons for telling these stories before we get into the details of what is possible at the frontiers of science today. The first is that we need to acknowledge scientists' tendency to push the doors of what is possible; the reactions of the society around them are what keep these tendencies in check. The second is to show that science does not occur in a vacuum. Those who push it forward often do so for a reason, and those who decide what should be done often have less than altruistic motivations. The first forays into blood transfusion were almost jingoistic: excessive interest in national pride overrode any sense of scientific purpose. The lure of funds from the American Association for the Advancement of Atheism was a significant driver for Ivanov; subsequent research funds from the Soviet state were for similar ends: it was widely believed that an ape–human hybrid would crush lingering religious feelings in the peasant population. There is no evidence that anyone involved in the scientific procedures gave much thought to concerns about what their patrons might have hoped to achieve by owning an ape–human hybrid.

It is also worth pointing out that the reporters of such research put their own spin on events. Rossiianov was disgusted by what he found out about Ivanov. 'I dare say, I found them disgustful,' he told *Outside* reporter Jon Cohen in 2007. 'Even now I find it terrible difficult to understand.' Two bioethicists who wrote an influential (and highly informative) book on the debate over human–animal chimeras are similarly partisan when one examines their writing in detail. David Albert Jones and Calum MacKellar repeat the canard, for instance, that Joseph Stalin ordered Ivanov to create ape–human hybrids in order to restore the Soviet war machine to full strength. 'According to Moscow newspapers,' they say, 'Stalin told the scientist: "I want a new invincible human being, insensitive to pain, resistant and indifferent about the quality of food they eat."' They support the argument by pointing out that 'Soviet authorities were indeed struggling at that time to rebuild the Red Army after a series of bruising wars.' However, nothing about Stalin's diktat is mentioned in Rossiianov's paper; the 'order from Stalin' is only traceable to a 2005 *Scotsman* article.

Jones and MacKellar suggest that there is little medical reason to pursue human–animal chimeras. Is that also false? It's an important question because we are now living in the age when human–animal chimeras, or 'animals containing human material' are a firm reality.

You could argue that we are already chimeras: we are, after all, a mixture of human and bacterial cells. Every day your several trillion human cells are carrying around several hundred trillion bacteria – an extra kilogram of alien biology – without which you could barely function. Your immune system relies on signals from bacteria living in your gut to alert it to dangerous intruders. Your outside is also coated with bacteria, somewhere around a thousand different species, according to an analysis published in the journal *Science* in 2012. These, too, exchange signals with the immune system to help you fight off infections.

If we are mostly bacteria, with a human centre, we have had our revenge. In pharmaceutical factories around the world there are vatfuls of bacteria that have a biologically human centre. All they get to do, all day, every day, is produce human insulin for the 35 million people with type 1 diabetes. These sufferers are unfortunate enough to have a body that doesn't create the protein that enables their cells to access the energy provided by their food. It's relatively easy to overcome the problem: they can inject the protein – insulin – into the fat beneath their skin, from where their circulating blood will take it where it is needed. Where do diabetics that need insulin get it? From chimeras: bacteria engineered with a human gene that allows them to take in nutrients and excrete insulin.

Strictly speaking, the insulin-producing bacteria are transgenic chimeras: that is, they contain genetic materials from another species, but not whole cells. There are numerous different types of chimera. The 'true chimeras' are composed of a patchwork of cells from two different animals that may be of the same or different species. These can be created by two embryos fusing, either in the womb or in laboratory experiments, or by injecting a developing embryo with stem cells from another animal. Then there's the 'true hybrids' – the kind of thing that

Ivanov was so keen to create. These result from the mixing of egg and sperm from different species; Ivanov's success breeding horses with zebras shows this can, on occasions, create successful embryos. In case you think the animals involved have to look roughly alike, it's worth pointing out that thousands of hamster eggs have been fertilised by human sperm. Not in their natural state: the eggs' zona pellucida, the protective outer layer that is normally only penetrable by a sperm from the same species, is etched away with chemicals. After that it can be used as a fertility test for men: the 'hamster test' shows whether their sperm have enough motility and power to penetrate the outer wall of a hamster egg.

The hamster test has largely been supplanted by newer fertility tests that examine sperm directly. It was never legal to keep the resulting embryo (if that is really what it was), and certainly not to implant it in a uterus of any species. But creating such a true human-hamster hybrid remains a possibility.

Next up is the cell hybrid, where two cells from two different species are fused together. These aren't sperm or egg cells, and can't develop into animals. They can, however, be cultured into a set of cells where each one has one set of chromosomes (packaged genes) from each species. Researchers might fuse a series of rat and human skin cells, for instance, while altering the chromosomes present in the human cells. Watching what chemicals the resulting cells produce can reveal what the different human genes do.

Cybrids are different again: the term is short for cytoplasmic hybrid and refers to a creature resulting from the transfer of a donor's DNA into an egg from another species (the egg's original DNA has been removed). A cybrid can also be made when a cell nucleus is implanted into a cell where the original nucleus has been taken out. The DNA donor takes over, with the egg donor contributing only around 0.1 per cent of the final creature.

When human DNA is involved in a cybrid, internationally agreed legislation says the embryo must be destroyed within a few days. That's enough time to harvest stem cells from the growing embryo, however – which makes it useful for medical research. That, fundamentally, is the point of much of the human chimera studies: to have something in a Petri dish that can help us understand where and why various human ailments arise. Or that's the hope – and the justification.

Creating human chimeras, by putting human cells into animals, often opens up the opportunity to make the animal susceptible to a uniquely human disease – which then allows it to be studied in ways that would be considered unethical in human studies. Putting human genes into animal eggs, for instance, is 'almost routine' now, according to a report from the Academy of Medical Sciences. It is used to investigate diseases such as Huntington's disease and muscular dystrophy. We have spent more than half a century grafting human cancers into mice to study how they develop. Chimeric mice with humanised immune systems have played a central role in the race to beat HIV. Rats with grafts of human neural stem cells help us understand how to repair brain damage after a stroke. And then there's the hope of chimeric organs.

If you want a pig to grow a human pancreas, the first step is to take a pig cell nucleus and knock out the gene responsible for growing a pig pancreas. That gene is called PDX-1. Once you've done that,then you take a pig egg and remove the nucleus, which contains all the egg donor's genes. In place of the nucleus, you put your pig nucleus with the knocked-out pancreas gene.

If this egg were allowed to develop like this (and it were able to develop successfully), you'd get a pig in a poke: life without a pancreas is not tenable. But, Townes and Ryan reasoned, if you introduce human pluripotent stem cells – the cells that can become any cell in the body – they will fill the gap. The full-grown pig would have a pancreas after all: a human pancreas.

If that sounds like science fiction, it's not – far from it. In 2010, when Sally Slater was celebrating the tenth anniversary of her life-saving transplant, Hiromitsu Nakaushi and his colleagues in Tokyo and London were celebrating an extraordinary success with their work. They had grown a rat pancreas inside a mouse by mixing pluripotent rat stem cells into a mouse blastocyst (a blastocyst is an embryo that has been allowed to develop a little). The mouse blastocyst had had its PDX-1 gene knocked out; while the rat cells mixed in with the mouse cells, making every part of the mouse a little ratty, the pancreas was almost entirely made of rat cells. It was also healthy and fully functional. The mouse was fine – until it was euthanised to have its pancreas inspected, that is.

That is why in June 2013 the Japanese government made it legal for their researchers to put human stem cells into animals. Until then Japanese law was, like the law everywhere else in the world, holding up the development of organ farms. But not everywhere else in the world had a researcher who knew how to farm in this way. Remember how miffed the British were when the French performed the first interspecies blood transfusion? The Japanese were not going to risk that kind of humiliation when their researchers had done all the ground work.

By June the Japanese had a new success and on 3 July researchers from Yokohama City University unveiled the mouse with a human liver. It was a proof-of-concept that may not reach medical applications for another decade, but it was still an astonishing breakthrough.

The liver was grown from human stem cells – taken from skin – that had been chemically reprogrammed to wind back their clock and return to an existence as pluripotent stem cells. More chemical treatments induced them to turn into the various kinds of cells found in the liver. Once these 'liver buds' were transplanted into a mouse, they connected up to the mouse's blood supply and continued to develop. Eventually, the scientists fed the mouse chemicals that break down differently in mouse and human livers. The breakdown products, sampled from the mouse's bloodstream, were those that are created in human livers.

It's not quite enough to knock out the pancreas-building gene, however. Not all of the Nakaushi mouse's pancreas cells were made of rat, because the arteries and blood vessels that supply the pancreas are built from instructions found in another gene. That's a hurdle that Nakaushi has overcome, though. He has knocked out the genes responsible for building blood vessels and arteries in mice. The pluripotent stem cells fill in the blanks and create rat-tissue networks for carrying blood all over the body. Everything is in place for the grand finale.

The idea, you won't be surprised to learn, is to grow a whole human liver. That can't be done inside a mouse, which is why Nakaushi has repeated the mouse/rat technique in pigs, which are the right size to host human organs. So far he has grown one pig's pancreas inside another pig. He plans to create a pig with human arteries and blood vessels and pancreas very soon. There's just one hitch: he's got to make sure that the pluripotent human stem cells don't turn into the wrong kind of cell. If they become egg or sperm cells, or brain cells, Nakaushi will have broken a law that even chimera researchers fear to transgress.

MacKellar and Jones describe the first potential catastrophe of the chimera age with aplomb. It's possible, they say, that 'some human cells might find their way to the developing testes or ovaries, where they might grow into human sperm and eggs. If two such chimeric mice were to mate, a human embryo might form which would be trapped inside a mouse.'

You can't imagine it, can you? We are simply not able to think in terms where a mouse is being swollen up from the inside by a human forming in its womb. It's highly unlikely that such a thing is even possible: the nutrients and infrastructure to support a human embryo almost certainly aren't available inside a mouse womb. Nonetheless, the truth about a mouse's ability to conceive a human is not something anyone wants to find out.

If Mackellar and Jones gave us the nightmare scenario, the UK's Academy of Medical Sciences raised a more plausible one. When the Academy issued its report on chimeras – what it termed 'animals containing human material' – it made clear that half-animal, half-human embryos – the kind of thing that Ivanov was trying to create – are now a possibility; 'some work does result in the presence of functional human sperm and/or egg cells in animals – which raises the remote possibility that fertilisation between human and animal germ cells might inadvertently occur.'

It's still unlikely, of course. If it were to happen, a viable embryo is even more unlikely. Nonetheless, the Academy recommended that a national body of experts should give the scenario careful consideration before any scientists are allowed to make that ultra-remote possibility possible. Nakaushi says he has found a way to stop the pluripotent stem cells from developing into sperm or egg. Are we willing to take his word for it? The same question might be asked about his claimed ability to stop stem cells developing in the brain.

This is the other nightmare: the pig – or monkey or mouse – with a human brain. This wouldn't even be an accident – not necessarily. One of the research goals for human–animal chimera work is to introduce human neural stem cells into primate brains in order to investigate how Parkinson's disease inflicts its misery. 'The key question, which cannot at present be answered with any certainty, is whether populating an animal's brain with human cells could result in that animal developing some elements of human consciousness, or "human-like" behaviour and awareness,' says the Academy.

It sounds ridiculous – laughable, as some experts have put it. But there are odd little precedents that should give us pause. Something else that sounds ridiculous and laughable is that, at Harvard University, with financial support from the National Institute of Mental Health, Evan Balaban painted the beaks of chickens with luminous T-shirt paint. But the result was no joke – it is regularly cited as reason to be very cautious about where we allow human brain tissue to end up.

Balaban had performed surgery on these chickens while they were still in development. He took some brain cells from Japanese quail embryos, cut tiny holes in the chicken eggs and implanted the quail cells in the chicken embryos' brains.

When the chicks hatched, Balaban took them to a recording studio he had prepared at Harvard and videoed their behaviour. The luminous paint meant that the chicks' head movements could be easily picked up by the camera. That was important because, as they crowed, the chicks bobbed their heads as if they were quail. Their crowing was also that of quail. The implanted cells had taken over.

If it crows like a quail, and moves like a quail, that doesn't mean it is a quail. Balaban's animals were unmistakably chickens. But many of their behaviours were unmistakably quail-like. Hence the concern over finding that we have created animals that contain human brain cells. How many does a mouse need before it displays human behaviours, say?

It's a question that becomes all the more pertinent as we seek to understand the roots of Parkinson's and Alzheimer's diseases; one idea in the mix is to study the brains of great apes that have received graftings of human brain cells. Not everyone is convinced this is a good idea, however. A group of researchers writing in the journal *Science* in 2005 suggested this could alter the moral status of a monkey. If our graft changes the primate's ability to feel pleasure and pain, to use language and rationality, to enjoy richer relationships, everything could change. 'To the extent that a [non-human primate] attains those capacities, that creature must be held in correspondingly high moral standing,' they argued.

Of course, this depends on the assumption, which we have already questioned, that the natural animal shouldn't already have that 'high moral standing'. But before we get to that, let's pause and reflect on what we already feel about chimeras.

Research into chimeras is promising, but nothing is guaranteed: there is no way to tell whether it will prove fruitful in the long term. Meeting demand for human organs would involve slaughtering a million pigs per year if we were to start farming pigs with a new slant on practices relating to organs. No matter how much you love bacon and sausages, that's a lot of pigs – enough to make anyone slightly queasy. On the other hand, imagine the grief of Sally Slater's parents had no donor come over the horizon. Which is worse? It's a conversation some people have started, and it's already clear there are no easy answers.

Take the 1996 report on the issues surrounding xenotransplantation – humans receiving organs from other animals – for example. Carried out by the Nuffield Council on Bioethics, the report asked for public reactions. It got them in spades.

'One cause of unease is the breaching of normally inviolate boundaries,' one respondent said. '[To] receive an organ from another animal might be seen as a mixing of one's human essence with that of the animal, and therefore as a dilution of one's humanity.' The report highlighted concerns about the reaction of those who see some animals as ritually unclean. First in line for xenotransplanters is the pig, whose organs are roughly equivalent to ours in size and function. But both Jews and Muslims might hesitate to accept such a transplant.

It's not just about religion. A small Australian survey of acute-care nurses found that two-thirds would not accept an organ from a pig. A similar proportion of potential recipients in the UK expressed reservations.

The Nuffield report concluded that, despite all the concerns, there were good medical reasons to press ahead with research into xenotransplantation as long as strong ethical and regulatory frameworks were in place. For some, such frameworks are not enough. In 2008 the BBVA Foundation, a Spanish scientific society, asked 22,500 people a similarly difficult question: 'To what extent do you think it is morally acceptable or not to create hybrid embryos a few days old to obtain stem cells that would be used solely and exclusively for research purposes?'

The respondents came from twelve European nations, Japan, Israel and the United States. They were told what exactly was involved: a nucleus is extracted from the egg of an animal (a rabbit, or a cow for example) and replaced by the nucleus of an adult human cell. The stem cells obtained from this hybrid embryo that is a few days old will be practically 100 per cent human. These stem cells could only be used to advance biomedical research. They would never be used in patients and nor would the hybrid embryos be allowed to develop beyond this preliminary stage. They would certainly not be implanted in patients.

The response was predominantly negative. Across the board it was not considered morally acceptable to do advance biomedical research by mixing human and animal material. Follow-up questions determined why: there was fear that breaking the species barrier would create novel opportunities for bad things to happen.

Others, though, can be positive about the opportunities. 'I have friends with MS and epilepsy,' said a respondent to researchers putting together that report for the UK's Academy of Medical Sciences. 'They are still alive thanks to the drugs they are taking. I am glad to still have them around me. I love animals to bits, but we have to move forward in medicine, saving the lives of our families.'

Here's another view from the same pool: 'I don't necessarily like it, but if it delivers benefits then it's worth it.' But then you get this, from someone talking about the possibility of animals carrying human reproductive cells: 'That is so far out there, just awful. Perhaps if there was no sperm left on earth, but otherwise no way.'

Where do you stand? You might think it doesn't matter – scientists will press on and do what they do – but that is not the case. The role of external opinion has long been underestimated in the direction science takes. Sometimes it holds science back, an influence that can delay a better world or prevent a worse one. Sometimes it offers a steer, taking us to places the scientists never envisaged.

If we are unsure about chimeras, epigenetics is similarly difficult. In many ways, this one is even more complex: we are about to find science beset by issues of politics, war and the burden of history, with its findings pounced upon as justifications for implementing policies and claiming reparations. Which is very confusing when you're still uncertain about exactly what the emerging discoveries in epigenetics really mean.

4

THE GENE GENIE

There's more to life than DNA

*From athletics to obesity, and from genius to crime,
the more we learn about genes, the more important the
environment appears to be.*

Steve Jones, The Serpent's Promise

It is not an obsession, but I am an interested observer of my genetic inheritance. That's how I ended up standing in the street in front of 772 St Nicholas Avenue in Harlem, New York. It's a brownstone building, big and solid and just like the others in the row. Though this was my first time in Harlem and I was a little uneasy standing in the street and staring up at a building, I felt a tiny sense of belonging. This was where my great-grandmother lived.

Laura Brooks was her name. I have a photograph of her: she is sitting on an armchair at the foot of a large staircase, a smile playing in the corner of her mouth as if the photographer has just told her a terrible joke. She was a seamstress, and earned a good living by all accounts – good enough that when she sailed to Edinburgh for a two-week visit to her son, my grandfather, she bought passage in her own rooms on the *Queen Mary*. She was not going to travel in steerage just because she was a negro.

I never knew her. Nor any of my great-grandparents, nor my grandparents on my father's side. I first met my father when I was twenty-one; he had left home when I was just one year old. People often ask me what that first meeting was like. The overwhelming memory for me is of lots of things suddenly making sense. We shared the same body shape. We seemed to laugh at the same kinds of jokes; I remember thinking that sense of humour must be a genetic trait.

Since that meeting I have been slowly piecing together my past. I am unaccountably pleased at the discovery that I have Jamaican genes: my children and I felt an absurd pride when we watched Usain Bolt win the 100 metres at the 2012 Olympic Games in London. Especially as I had told them that their great-grandfather – my father's father – was a sprinter on the track team at George Washington High School in New York. We Brookses are powerfully built, I said: it's an upside of being the descendants of Jamaican slaves. It's only fair, because there are certainly downsides.

'It is now broadly known that an African American man in Harlem is less likely than a man in Bangladesh to survive to the age of 65.' It's an appalling statistic, and when Christopher Kuzawa and Elizabeth Sweet used it in the opening paragraph of a 2009 paper on the emerging science of epigenetics, they knew it would appal people. They continue the assault:

> Nationally, African Americans have an age-adjusted all-cause mortality rate that is 1.5 times that of whites, and cardiovascular diseases (CVD) and their precursor conditions, including

hypertension, diabetes, and obesity, contribute heavily to this disparity. The risk of dying from heart disease is 1.3 times higher in African Americans compared to US whites, and African Americans are 1.8 times more likely to develop diabetes. Hypertension rates are roughly 1.5–2 times higher in African Americans compared to whites, and are especially high in certain regions, such as the so-called 'stroke belt' of the American South. In total, nearly half of all African American adults develop some form of CVD, making racial disparities one of the most pressing US public health problems today.

In the 1980s researchers at Southampton University who were trying to find what lay behind such statistics hit upon an interesting correlation. Diseases such as cardiovascular disease, diabetes and hypertension were all more common in adults that had been born at a weight below the average. There was a clear link between low birth weight and a shortened life expectancy. The reason for this was eventually pinned on the notion that a foetus experiencing deprivation would allocate resources differently from one blessed with adequate nutrition. The most obvious adjustment would be slower growth, but organs and other aspects of physiology might be tweaked. One well-documented example is the fact that children undernourished in the womb have much smaller kidneys – and a consequently higher chance of renal failure in later life.

What's more, the damage is permanent. Give a low birthweight child a high-nutrition catchup diet after birth, and it will still be subject to a greater risk of the constellation of diseases associated with a low birth weight. Malnourishment in the womb is a permanent handicap. It's just one reason why we need to get to grips with the science of epigenetics.

'Epi' means over or in addition to. If you want to know how an embryo – or anything else biological – will develop, it's not enough to know about the genes, the chemical units that play a role in determining eye and hair colour and a host of other traits and characteristics. You also need to know about the environment in which those genes' chemical properties are operating. As geneticist Steve Jones says in his book *The Serpent's Promise*, 'The more we learn about genes, the more important the environment appears to be.'

Let's start, though, with the genes. The study of inheritance began in earnest with the Augustinian friar Gregor Mendel. Mendel joined the priesthood for financial reasons, was hopeless at dealing with parishioners and failed his exams to become a science teacher. Almost all that was left to him was a quiet life in a monastery in the Augustinian Abbey of St Thomas in Brno. There, however, he undertook an inspired project to discover the laws of heredity.

'It requires indeed some courage to undertake a labour of such far-reaching extent,' Mendel wrote in the 1866 presentation of his results. When he embarked upon the experiment, the received wisdom was that we inherit a blend of our parents' characteristics, with everything averaging out. Mendel's experiments with 29,000 pea plants blew that fusty old idea away. When he hybridised pure-bred plants with white and purple flowers, for instance, he found that the next generation had purple flowers. From experiments such as this, he guessed that there existed units of heredity – that an instruction to 'be purple' must somehow dominate over the instruction to 'be white'. Eventually, he used the experimental results, his knack with logic and mathematics (and a little sleight of hand) to distil his findings down to two laws.

One was the Law of Segregation. For any trait, such as colour, a plant possesses two factors, one of which may be dominant over the other. Only one of these is passed on to the next generation; the daughter plant gets the other factor from the other parent.

The dominant factor determines the physical appearance of the daughter.

The second law was the Inheritance Law. This said, put simply, that each factor was independent of other factors. Two different physical characteristics, such as leaf shape and petal colour, are passed on by two different factors.

That was it. It took him eight years, and experts agree that he must have cheated. His experiments apparently produced generations of flowers with physical characteristics that were in perfect – too perfect – agreement with the laws of heredity. Maybe that explains the final sentence of his introduction to the work: 'Whether the plan upon which the separate experiments were conducted and carried out was the best suited to attain the desired end is left to the friendly decision of the reader.'

Only years after his death did anyone appreciate that Mendel had identified something profound. The delay wasn't due to doubts over his data; the problem was that Mendel had committed the cardinal sin of separating what was inherited from what developed. Mendel's work said nothing about the mechanisms behind the way the physical traits manifested in the new plants. At the time no one was researching inheritance without reference to development, and so no one took any notice of the blatherings of a German monk. That's why, three decades later, in the years after Charles Darwin had published *On the Origin of Species* and posited the need for a unit of inheritance, a collection of researchers reinvented the wheel, discovering Mendel's laws for a second time.

Darwin's theory of evolution, coupled with a new, mathematical approach to heredity championed by his cousin Francis Galton, was a powerful fillip for biology. In 1894 Cambridge University researcher William Bateson published a book of observations of natural variation, calling for 'the organization of systematic experiments in breeding'. Soon after it came out he found that Mendel had already done it. Bateson translated Mendel's works into English and, in 1905, coined the term 'genetics' from the Greek word for giving birth. The effort to distil the unit of inheritance had begun.

Genes, we now know, are large molecules that contain the information needed to create the proteins that do the heavy lifting of biology. Our genome has four types of molecule, represented by the letters A, T, C and G, the alphabet of the genome. Its words are three letters long, and grouped into paragraphs known as exons. Paragraphs are lumped together to make genes. These are assembled on long strands of sugar and phosphate molecules, with the A,T, C and G molecules running sideways off the strands. The way these letters are combined creates a vast instruction book that is read by other molecules in the body. Those molecules use the instructions to make proteins.

When we first began to isolate genes, it was clear that there were differences between the DNA strands that appeared in different organisms. Where one has a G, another might have a C. Where one has a T, another might have a C. The number of different possibilities is enormous, which is why we can have such an enormous variety of animals and plants, including great physical variety within species.

It's a beguiling discovery that promises to explain so much. When the human genome project to find the sequence of letters that provides the recipe for humans neared completion, the geneticist, author and pathological optimist Matt Ridley wrote in his book *Genome*, 'In just a few short years we will have moved from knowing almost nothing about our genes to knowing everything.' Ridley was not alone in his hyperbole. In

2000, just before the sequence of the human genome was published, the journal *Nature* called the Internet gateway to the data the 'Golden Path'. In 2010, in the journal *Cell*, researchers using the ability to mine a genome for evolutionary data suggested we were now in a 'golden age for evolutionary genetics'. According to a 2011 paper in the journal *Science*, we are now living in the 'Golden Age of Human Population Genetics'. Titles of other publications include 'The Yeast Genome: on the Road to the Golden Age'; 'A Golden Age for Microbial Genomics'; 'Genetics and the Golden Age of Biology' … I could go on, but I won't, because it's all far from true.

The truth is, we have known for decades that it's not just about genetics. Analysing the genomes of organisms from *E. Coli* bacteria to gorillas to (golden) retrievers promises a great deal. It tells us the sequence of millions of genes, each one a recipe for making a protein that performs some biological function, that encodes the organism. But you can have an identical set of letters in the DNA code, and still get organisms that are different from one another. That's because genes don't just exist, or not exist. They are turned on and turned off by chemicals in their environment. And those environmental factors can have consequences – often profound and distressing consequences – that last for generations.

The first glimpse of the power of epigenetic changes came in the early 1990s. That was when genetics researchers put the spotlight to the descendants of the humanitarian catastrophe that produced the delicate, elfin features of Audrey Hepburn.

Hepburn is best known for her role as Holly Golightly in the film of *Breakfast at Tiffany's*. Not that that's her character's real name: it turns out the Manhattan socialite is really a Texan girl called Lula Mae Barnes who got married at fourteen and ran away, hoping to find a way to support her brother when he comes out of the army.

The script must have resonated for Hepburn, herself a reinvented fugitive from a disastrous past. In the winter of 1944 she was a teenage girl who called herself Edda van Heemstra. She lived in German-occupied Arnhem, in the Netherlands, and her British-tinged birth name, Audrey Kathleen Ruston, would have raised suspicion.

Not that she was without courage: she performed secret ballet recitals to raise money for the Dutch resistance. The performances must have been a challenge: like most of her audience, Edda was suffering anaemia, dropsy and respiratory problems. The entire population of the region was being systematically starved by the German occupying force as a punishment for refusing to help fight the Allies. The period was known as the Dutch Hunger Winter and it saw the deaths of 22,000 people. It would affect Edda for the rest of her life. When the Allies liberated Arnhem she had added jaundice, endometriosis, asthma and depression to her other illnesses. She changed her name back to Audrey, but she could not change the legacy of her privations.

Studies of victims of the Dutch Hunger Winter have revealed several remarkable facts. Girls born to women who were in the early stages of pregnancy during the famine were twice as prone to developing schizophrenia. We already know there is a genetic component to this disease – you are more likely to suffer from schizophrenia if your parents or other close relatives do. The relationship is complex, most likely involving thousands of genetic components, but something was happening to the genes of the babies in the womb of those mothers.

Another surprising outcome was a tendency to obesity. Mothers malnourished early in pregnancy produced babies that were more likely to become obese in later life. Again, genetic changes are implicated – we know that a tendency to obesity is partially heritable. Mothers who were malnourished late in pregnancy – when the baby is doing most of its growing – gave birth to infants of significantly lower weight. These babies were never able to catch up, staying smaller than the average population and with a reduced likelihood of becoming obese.

Women born during the Dutch famine were more likely to develop breast cancer. The children of the famine are also twice as likely to develop coronary heart disease. These too are diseases known to have a hereditary factor. Malnutrition in pregnancy seems to have a long-lasting effect on the genetics of your child.

And on your grandchildren, as it turns out. Researchers are still only beginning to tease out the effects that cascade down the generations whose forbears are exposed to environmental stresses, but results from studies of the Dutch famine victims' descendants suggest that the alterations to which genes are turned on or off survive at least two generations: the one that suffered in the womb during the famine, and their children. They may go much further.

We are all aware of the effect of genetics on hair colour. It is particularly remarkable in redheads, where a single gene – it's called MC1R – carries the responsibility for promoting the expression of red hair. Agouti mice, however, can change the colour of their offspring's fur while their genome stays unchanged. All you have to do is give a pregnant female the right kind of food.

In 2000 Randy Jirtle and Robert Waterland fed a pregnant agouti mouse on a highly nutritious diet rich in a group of compounds known as methyl donors. The name comes from the chemistry of methyl groups: a carbon atom attached to three hydrogen atoms. Methyl donors are contained within foods that allow this group of four atoms to detach easily from the other molecules and travel through the body. Ultimately this unit of carbon and hydrogen atoms attaches itself to our DNA, determining just how active our genes are. Onions, beets and soy milk are strong methyl donors, as are the folic acid supplements that pregnant women are advised to take.

The pregnant mouse had yellow fur. Her offspring had brown fur. The reason? Methyl groups from the foods had made their way into the chromosomes of the embryonic mice and shut off a gene known as the agouti gene. With that gene shut down, the hair follicles produce brown fur.

Such methylation is not normally passed on to the next generation. In the gametes – the sperm and eggs – of brown agouti mice, the methyl groups are mostly removed. Should they get through, another reset routine does another comb-through of the chromosomes, looking for methyl groups to strip away. But some do survive both attempts at purging. As a result the diet of a pregnant agouti mouse during gestation can affect the coat colour of her offspring's offspring – and the effect can tumble on down through succeeding generations.

It's an important finding because the agouti gene is not just about coat colour. It's also about health. Yellow agouti mice tend to be fat. They have a ravenous appetite and are more prone to cancer and diabetes than their brown, genetically identical counterparts. That's true of mother and yellow baby – and, it seems, that baby's yellow baby. This transgenerational epigenetic inheritance of an increased chance of poor health makes it very interesting to ask whether epigenetic changes can affect humans in the same way.

It seems they can. Though it is very difficult to prove that a problem is caused by an epigenetic change rather than a straightforward genetic variation, there are some indications that epigenetics can be a killer inheritance. In 2007, for instance, Australian researcher Megan Hitchins found a case of a hereditary colorectal cancer marked out by epigenetics. One of the patient's 'DNA mismatch repair genes' had been methylated, rendering it inactive. The way it was silenced suggested a parent had passed on the epimutation. In 2003 Karin Buiting's group in Essen, Germany, found that silenced genes responsible for the onset of Prader-Willi and Angelman syndrome had been methylated in their grandparents' generation – if not before. And before all that was the life-changing work kicked off by Yellapragada Subbarao.

You have almost certainly never heard of Subbarao, and that's the way he would have wanted it. He was born in 1895 in Andhra Pradesh in India to a mother who sold her jewellery to pay for his education. Subbarao repaid the debt by becoming a doctor and medical researcher who discovered a cure for tropical sprue, the disease that had killed two of his brothers and almost killed him.

Tropical sprue results from an imbalance in the gut bacteria. It is most often seen in tropical areas. Sufferers can't absorb nutrients properly and waste away. The cure is fairly straightforward: administration of tetracycline antibiotics, folic acid and vitamin B12. Subbarao was the first to bring us two of these: he was instrumental in the discovery of the first tetracycline 'broad spectrum' antibiotic, and he was the first to isolate folic acid.

Subbarao had a remarkable facility for biochemistry. As well as finding our most powerful antibiotic, he created and tested aminopterin, the first anti-cancer drug. He also discovered the fuel used by almost all biological systems: adenosine triphosphate or ATP. For most people that might have been enough to put him under the noses of the Nobel Prize committee. But Subbarao showed little interest in collecting awards or accolades – or money for his medical inventions; a patent lawyer once dismissed him as 'a poor businessman'. It's clear that Subbarao had higher things on his agenda. After he died in 1948, his colleagues announced that his last declared wish had been, 'If God will spare me another couple of years, may be we can cure another disease.' He wasn't spared, but his work did show us how to cure another disease. Within a few decades it was clear that his isolation of folic acid would be the answer to neural tube defects in infants.

If it forms correctly, the neural tube encloses the spinal cord and the brain. It starts as a groove that deepens, then closes off. However, a folic acid deficiency can prevent that closing process, leaving nerves exposed. That results in a range of conditions, the best known of which is spina bifida. In the worst cases of neural tube defects, the brain doesn't form properly, creating a disturbing condition called anencephaly.

Folic acid, it turns out, provides a methyl group – that unit composed of a carbon atom and three hydrogen atoms – that binds to foetal DNA and ensures that genetic instructions are carried out properly, fully closing the neural tube. The most definitive proof came via a project known as the 'MRC Vitamin Study'. Sanctioned by the UK's Medical Research Council, despite claims that it was unethical, it looked at the effect of various nutritional supplements on the pregnancies of two thousand women known to have an elevated risk of having babies with a neural tube defect. It launched in July 1983; by April 1991 it was considered unethical to carry on. The results were so stark that it would have been immoral to allow any more pregnant women to go without folic acid supplements. Folic acid was preventing somewhere between 70 and 80 per cent of neural tube defects.

A number of countries took the results and ran with them. The US, Canada and Oman, for instance, mandated that flour and cereals should be spiked with folic acid so as to ensure anyone at risk of falling pregnant would have enough folic acid in their body to grow a foetus with a markedly reduced chance of a neural tube defect. Today more than sixty countries do the same. Lamentably Britain, where the original study was carried out, is not among them. That's why various British charities and the National Health Service encourage any women who are sexually active to take a folic acid supplement every day. The first four weeks are a crucial time in the development of an embryo's brain and spinal cord; if you fall pregnant, a lack of folic acid in your body could cause neural tube defects before you even know you are pregnant.

It is hard to believe that something so catastrophic can be so easily preventable. This is where we see the limits of concerning ourselves only with the genome, and the crucial importance of accounting for epigenetics. The reason this revolution has been so long coming can be traced back to the unfairly sullied reputation of a Frenchman: Jean-Baptiste Lamarck.

Poor Lamarck. Charles Darwin called him a 'celebrated naturalist' and noted that he 'first did the eminent service of arousing attention to the probability of all changes in the organic, as well as in the inorganic world, being the result of law, and not of miraculous interposition.' But although he came up with a perfectly respectable and reasonable theory of biological evolution, no one ever acknowledged it as a useful scientific contribution during his lifetime. When his place of work was rebranded as the National Museum of Natural History after the French Revolution, Lamarck was given the lowest of the new professorships: Professor of Worms. He died blind and penniless, and was buried in a rented space in a lime-pit, then exhumed and thrown into the Paris catacombs. No one knows where his remains now lie.

Lamarck, like Mendel, saw two laws of inheritance. One was the inheritance of acquired traits. A wading bird would stretch its legs and spread its toes in order to make more of the feeding opportunities, for example; this would eventually result in long legs and webbed feet that are passed on to its offspring. Perhaps the most famous example is the notion that early giraffes had short necks. Generations of giraffes developed stretched necks as they reached for leaves on high branches and passed the characteristic stretched neck to their offspring, so ultimately all giraffes came to have long necks. Conversely, little-used traits would disappear from the lineage.

It sounds vaguely absurd, rather like Kipling's *Just So Stories* – how the leopard got its spots, and so on. But in a time when no one could explain how the variety of life had arisen, it was a perfectly reasonable hypothesis. Indeed, part of the rancour against Lamarck came from his writing God out of the deity's assumed role of creating every tiny speck of nature's variety. In many ways Lamarck was a very modern biologist.

Lamarck's second law was simply that organisms would gradually become more complex over time. This idea of progress, or direction, to evolution was anathema to those who came after. 'Heaven forfend me from Lamarck nonsense of "a tendency to progression"', Darwin wrote to botanist Joseph Hooker in 1844. The mud stuck. The idea of evolutionary 'progress' and the notion of the inheritance of acquired characteristics were ridiculed and reviled in equal measure for more than a century; in most centres of research they still are. Which makes it all the more courageous that, in 1940s Edinburgh, Conrad Waddington was willing to lay out his studies of the ostrich.

As it happens, many of the genes I carry could be found in 1940s Edinburgh. I have a photo of my grandfather dressed in highland garb, a kilt reaching to his black knees, pretending to play a set of bagpipes. His trip to a photographer's studio was close to his first act on arriving in Edinburgh from Harlem; he used the photograph as a postcard to announce his safe arrival to his mother back in New York.

He had come to study medicine at Edinburgh University (and, if rumours are true, to get away from an unpleasant situation involving a pregnancy). While he sat in lectures at the Department of Medicine, the field of epigenetics was slowly forming just down the street.

It was Waddington, the Buchanan Professor of Genetics, who came up with the name. Much of Waddington's early work was in the pursuit of the mechanisms by which an embryo organises its body plan. This is focused around the development of the 'primitive streak', which eventually gives rise to the neural tube – a structure whose correct formation is, as we have seen, highly dependent on epigenetic processes. Waddington became convinced that the search for genes was too simplistic. He conceived of something he called the 'epigenetic landscape', a panorama of hills and valleys that would exert a pull on a gene's instructions, causing it to create different products, depending upon the lie of the land at any particular point in the organism's development.

Besides this laying out of the scheme by which genes might be influenced, Waddington made only one other truly significant contribution. It is so significant because it rehabilitated – in part, at least – a long discredited notion: that animals could acquire characteristics during their lifetimes that would then be passed on to their offspring. The prime example, he said, was the ostrich.

When an ostrich sits down, there are two points on its body

where it will come into particularly forceful contact with the ground. At these two points, ostriches have callouses – areas of bare, hard skin where no feathers will grow. It would be safe, and sensible, to assume that these callouses arise during the ostrich's lifetime, growing a little every time it sits down on the dusty earth. However, the callouses are there when an ostrich chick hatches; they are formed during the development of the embryo.

How did the ostrich get its bare spots? It was born with them. It was impractical for Waddington to do experiments on ostriches, but he did discover similar Lamarckian traits in fruit flies. He learned how to breed flies such that a change in temperature during pupation would change the pattern of veins in their wings – a change that would survive for generations, even in the absence of temperature shock. In the right circumstances, a shock or stress changes a fruit fly's appearance. Lamarck may have chosen the wrong examples with which to illustrate his idea, but his idea wasn't entirely wrong.

Not that Waddington's idea was accepted without a fight. For decades, researchers tried to show that this was a genetic change – that it wasn't, as Waddington claimed in the title of his published paper, the 'inheritance of acquired characters'. No one wanted Lamarckism to be seen as even vaguely plausible, and especially not thanks to the work of someone as respected as Waddington.

It took until 2003 for the tide to turn. That was when a team led by epigeneticist Douglas Ruden of Wayne State University in Detroit, Michigan, showed that a compromise could be reached. Their research led to a view of the crossveinless flies as a 'passing on of acquired characteristics, unquestionably a Lamarckian concept'. Ruden's group tempered this heresy: 'importantly, those characteristics are still generated at random with respect to the needs of the organism, a Darwinian concept.'

To understand where that randomness comes in, we need to zoom in on the mass of DNA and surrounding proteins in a cell nucleus: an agglomeration known to biologists as chromatin.

A single strand of DNA can be anywhere from 1.5 to 8 centi- metres long; to fit this inside a cell nucleus requires some careful packing. That is done by proteins called histones, which wrap the DNA around themselves. Each histone takes a couple of turns of DNA, then passes the strand to the next histone. Once all the wrapping is done, the histones are pulled together into a spiral formation. The histone proteins are rife with sites at which other molecules can bind. Methylation is just one of the binding pos- sibilities. Histones that are supplemented by phosphate or acetyl groups, or other proteins, offer more than fifty different combi- nations of conditions that change 'expression' – that is, turning genes on or off, amplifying or quietening their instructions.

In the end the exact shape taken up by the histones and the bound-in epigenetic modifier molecules will affect which of the genes are accessible to the rest of the cell's molecular machinery. The enzymes that bind to the DNA in order to read and execute the genetic instructions might find a very limited set of access sites, thanks to the shape and the modifier molecules. If biology were a restaurant, DNA would be the menu, and the methyl groups, the histones and its bound molecules are the diners making the choices. It is they who determine which dishes are going to be cooked (by those enzymes) and which will remain nothing more than a possibility written out on the menu of options.

And, to continue the analogy, a variety of factors determine which diners will come to the restaurant. We have already seen that one is diet. Folic acid is crucial, but there are other sources of methyl groups; lots of soy, for example, means lots of methyl groups coursing through the body. Another is temperature: we know that cold temperatures induce epigenetic changes that give fruit flies a darker pigmentation. And then there are the curses of modern life: pollution and stress – modern epigenetic analysis shows these are more dangerous than we might ever have expected.

When Harvard School of Public Health's Andrea Bacarelli wanted to find somewhere pollution really might have changed the local epigenetics, he decided Boston wasn't extreme enough. Instead he went to Thailand; specifically, the Ma Ta Phut industrial estate in Rayong, a vast conglomeration of steel factories, oil refineries and petrochemical processing units. There Bacarelli's team analysed the blood of sixty seven plant workers and sixty-five local residents, and compared it with blood taken from people living in rural areas far away from chemical or industrial plants. The results were clear. Those living and working in the vicinity of the industrial estate had DNA that was much more methylated. Bacarelli performed a subsequent study of steelworkers that showed blood DNA methylation by airborne pollution – specifically small particulates – alters the way blood will clot. Another study, by Frank Gilliland's group at the University of Southern California, showed that air pollution seems to result in methylation of the gene responsible for nitric oxide production – known to be a cause of asthma and wheezing in children.

Then there are the results gained in 2011 at the Salk Institute in La Jolla, California. Researchers there showed that methyl groups can exert an influence on thousands of genes in an organism. That gives epigenetic effects hundreds of thousands of times the influence of spontaneous mutations, the only source of variation available in purely Darwinian thought. Darwin's thinking was that such mutations happened at random; we now know that this is not the case for epigenetic changes: they occur in hotspots on the genome, hotspots that are usually sites where they induce significant change to the organism's production of proteins. What's more, these changes last.

The Salk study showed that epigenetic mutations lasted for at least thirty generations. That's worrying when we learn the results of a Japanese study that showed how chemical or environmental stresses can cause DNA to unfold within the chromosomes. The unfolding exposes genes to activation (or silencing) by nearby molecules, and the unravelled DNA is passed on to the next generation. If those chromosomes happen to be in egg or sperm, enormous and potentially dangerous variations cascade down the generations. Expose one generation to hardship, and the next – and the next, and the next – will suffer.

It's a prognosis that is borne out by a study published in January 2013. A group of researchers from Cambridge University showed that, although standard biological thinking says that sperm and eggs are stripped of any epigenetic markers such as a methyl group, approximately 1 per cent of the changes get through the erasure process unscathed. That makes perfect sense. While the results of studies into the long-term effects of the Dutch hunger winter have been somewhat ambiguous, there's no ambiguity in a remarkable set of studies coming out of the barren wastes of northern Sweden.

Inside the Arctic circle, life exists on a knife edge. Occasion-ally, in snowbound Norbotten, the county at Sweden's northernmost tip, people have fallen from that blade. When the crops failed, as they did several times in the nineteenth century, there was widespread famine. There were years of plenty too, however.

We know this because the Swedes keep precise records of agricultural achievements. They also keep very informative records of births, illnesses and deaths – including details on the causes of death. Put all this together, as Lars Olov Bygren, a health scientist at Stockholm's Karolinska Institute, did, and you have a way of tracing any long-term effects of nutrition – and malnutrition.

Bygren's starting point was a sample of 99 people born in 1905. He traced their parents and grandparents, and noted how much food would have been available to them as a result of feast and famine years. Controlling for socioeconomic factors, a Swedish boy approaching puberty who over-ate in a glut year would risk reducing his grandson's life expectancy by thirty-two years. The punishment for their gluttony was dished out on their children's children.

At first Bygren's results were dismissed as ridiculous. It certainly does seem ridiculous, but it is borne out in other studies. If boys start smoking before the age of eleven, for instance – the age when their bodies are just starting to manufacture sperm – their sons will be significantly more overweight by age nine than their peers with fathers who only took up smoking later. The only way this can happen is if the act of smoking tobacco puts some kind of epigenetic marker on their sperm.

We are a long way from pure genetics now. Interestingly, though, we may be a lot closer to Darwin's original vision for his work.

Though he may have been the first negro to dress up as a highlander, my grandfather was far from being the first one to walk the streets of Edinburgh. During the centuries of slavery it was not uncommon to see freed slaves living out their lives in the cities of Britain. Edinburgh had more former slaves than most:

many of the plantation owners in the West Indies were Scottish, and they frequently returned with favoured servants in tow. That is how, almost a hundred years before my grandfather arrived in Edinburgh, a seventeen-year-old Charles Darwin came to spend many evenings in this city being taught how to stuff birds by a 'blackamoor' called John Edmonstone.

Edmonstone, a freed Guyanese slave, was working in the university museum, earning a guinea a term for giving taxidermy lessons. Darwin was in Edinburgh to study medicine, but he had no facility for it. The lectures bored him; the practical aspects turned his stomach. Taxidermy, and the company of Edmonstone, was of much greater interest to the young Darwin. He happily paid Edmonstone a guinea per hour, and took an hour's lesson every weekday for two months. They became 'intimate', Darwin later wrote; Edmonstone was a 'very pleasant and intelligent man'.

According to historians Adrian Desmond and James Moore, this was the first of a series of events that spurred Darwin to the theory of evolution by natural selection. Only in recent years, with new bundles of Darwin's correspondence coming to light, has it become clear that Darwin's aim in writing *On the Origin of Species* was to destroy all support for slavery and oppression.

Darwin came face to face with the reality of the slave trade during his five-year voyage on HMS *Beagle*. The ship set off from Plymouth just after Christmas in 1831. Britain had outlawed the trade in slaves in 1807, and part of the ship's mission was to patrol the South American coast for slavers. There was no shortage of such vessels: Darwin came across many loading up with manacles and branding irons, ready to sail for west Africa. His journal, published in 1845, records his encounters with 'heart-sickening atrocities'. When he saw a six-year-old slave boy whipped 'for having handed me a glass of water not quite clean', Darwin could contain himself no longer and stepped in to intervene. Most of the time, though, he felt there was nothing he could do. As a guest in one house, he witnessed a slave being beaten hourly. It was, he said, 'enough to break the spirit of the lowest animal'. Darwin records another occasion where he walked past a house in Brazil where he 'heard the most pitiable moans, and could not but suspect that some poor slave was being tortured, yet knew that I was as powerless as a child even to remonstrate.'

Darwin's aim in researching animal and human origins was to highlight our commonality, the brotherhood of man. He sought to show that the rich variety of organisms we see in the natural world could arise from a common origin, and that humankind was no different. As Desmond has written, Darwin's notebooks show his thinking moving 'from racial kinship and brotherhood to unite all suffering creation'. In May 1838 Darwin wrote, 'I cannot help thinking good analogy might be traced between relationship of all men now living & the classification of animals.' The notebooks tore a strip off the slaver, who, he said, 'has debased his Nature & violates every best instinctive feeling by making slave of his fellow black.'

It was a widely unpopular notion. The Christian church circles in which Darwin moved abhorred the idea of mutation. Their God would not create species in all their variety and splendour only to have them change, or create a dull worm and let it make its own way towards becoming a peacock. The groups profiting from slavery also hated Darwin's idea. Dehumanising the negro made it possible to treat the slaves like property, like animals to be worked into the ground and then discarded. The notion that they and the negro shared a bloodline made that attitude much harder to sustain. In their extraordinary book *Darwin's Sacred Cause*, Desmond and Moore demonstrate how Darwin campaigned through science, carefully gathering evidence to show the common ancestry of European and African men and women, and pulling the rug from under the pro-slavery arguments that slavery was a 'natural' state. *On the Origin of Species* was published in 1859. The United States abolished slavery in 1865. But despite Darwin's efforts, epigenetic studies such as Kuzawa and Sweet's suggest the legacy remains.

African-American children have always had a significantly lower birth weight than European Americans. This is not an issue of poverty; white women on low incomes have babies that are, on average, 200 grams heavier than those born to black women on equivalent incomes. The black women were better educated, smoked less, had higher body weight, lived in better housing and had healthier psychological profiles, but their babies were still much lighter. There's something in that maternal line: in children of mixed marriages, having a black father gives a higher birth weight than having a black mother.

Things have been this way since shortly after slavery, if not longer. A Johns Hopkins hospital survey of their data from 1897 to 1935 showed that black mothers gave birth to babies 7 per cent lighter than white mothers. In 1988 the difference was almost unchanged, 8.6 per cent. For black women, increasing birth weight is not about getting out of poverty. A 2006 study of inter-generational birth weights published in the *American Journal of Public Health* showed that white women are far less likely to give birth to a low-birthweight baby as the adult family income increases. For black women, increasing family income made no appreciable difference.

Researchers have observed this stubborn disparity for decades and applied every factor they could think of. Nothing has ever explained it. One study, carried out across the US in 1967, of 18,000 babies, concluded that race 'behaves as a real biological variable in its effect on birth weight. This effect of race [is] presumably genetic.'

The presumption was wrong: it isn't genetic. We know that because in 1997 two Chicago physicians compared the birth weights of over 90,000 infants born to US-born white women, Africa-born black women, and African-American women. The African-American population share three-quarters of their genetics with the Africa-born population. The other quarter is largely from European genetics. Nonetheless, white European-origin women and black Africa-born women had babies of almost identical birthweight, while the African-American women, the descendants of slaves taken from West Africa, had infants nearly 250 grams lighter than Africa-born women. If it isn't genetic, it's a long-lasting environmental effect. As Kuzawa and Sweet contend, we really have to stop obsessing over genetics and start paying attention to 'the more durable role that environments have on biology and health when experienced early in the lifecycle'.

Once we do start paying attention to epigenetics, we can begin to mitigate the effects that epigenetics has on the downtrodden, the poor, those living in polluted environments and those whose diets are adversely affecting their children down through the generations. Armed with this understanding, we can break the cycles that have ruined lives for decades – sometimes centuries. Epigenetics, it appears, will be the field that completes Darwin's efforts to make a better, more equitable world.

With the gift of hindsight, it almost seems obvious that diet and environmental factors can have such strong effects on the intricate molecular mechanisms of our bodies, and cause such radical effects on our health. We might say the same of our next subject: the role of gender in medical treatment. Men and women might not be from different planets, but we are certainly built in very different ways.

5

DIFFERENT FOR GIRLS

Men and women ail in very
different ways

Men are from Earth, women are from Earth. Deal with it.

George Carlin

In 1989 the World Health Organization (WHO) recommended
that a high dose of measles vaccine, known as the high-titer
measles vaccine, be used in countries where deaths from measles
were highest. It seemed like a good idea at the time.

Vaccines have been an extraordinary success story, saving millions of lives across the world. In many ways, we have become far
too familiar with vaccination: the idea of using a weakened virus,
or part of a virus, to train our bodies to resist a full infection was
a simple but brilliant innovation that has radically changed the
human experience for the better. However, it is so routine that we
take its benefits for granted. Almost no one in the West can remember what it is like to expect childhood disease to take at least one of
our children or siblings from us before their tenth birthday.

That said, vaccines are still something of an enigma. Getting
the vaccine dose, and the timing of its administration, right is

not always straightforward, for instance. In trials, increasing the dose of measles vaccine from 10,000 to 40,000 viral particles had increased the percentage of children with measles immunity from 73 per cent to 100 per cent. That was why, in 1989, the WHO recommended the use of 100,000 viral particle doses in countries where measles was a severe problem.

In the weeks and months immediately after the vaccine was administered, everything seemed fine. But as more time passed, reports coming from Haiti, Senegal, Gambia and Guinea-Bissau had made it clear that there was a serious problem.

In groups that had received a dose of the new high dose measles vaccine, 33 per cent more girls died than boys. They weren't dying from the vaccine directly; they were being killed by the standard diseases that afflict children in the developing world: diarrhoea, sepsis and infections, for example. But the girls were dying, nonetheless, and something had to be done. In February 1991 an international panel of experts were called to a meeting at WHO headquarters in Geneva, where they were presented with the results. The notion that this was due to the vaccine, the experts said, was implausible. However, in 1992 the WHO rescinded its recommendation and the high-titer measles vaccine was withdrawn.

No one could accuse the researchers involved of being high-handed or uncaring about the damage that was done in that short time: contrition certainly pervades Neal Halsey's account of his involvement in the story. Halsey is the Director of the Institute for Vaccine Safety at the Johns Hopkins Bloomberg School of Public Health, a hugely respected medical researcher, and a tireless campaigner for improved vaccine safety. 'In retrospect,' he says, 'I and other investigators were too easily convinced of the safety of high titer vaccines.' His retrospective, published in the *Pediatric Infectious Disease Journal* in 1993, is moving: he and others attempted to minimise the damage done by the vaccine. They couldn't take it out of the children's bodies; instead, they ensured that the girls it had put at risk were given extra food and improved access to medical care.

At the time, he says, he and his fellow researchers had had no reason to suspect that the vaccine, created from a weakened form of the measles virus, would be good for boys and yet bad for girls. More than twenty years later, we are still only scratching the surface of gender-based medicine.

On the face of it, the idea of gender medicine seems rather obvious. We all know men and women are different – it's one of the first things we learn as a child. There are obvious external differences, and they are coupled with clear internal differences: women have a cervix, for instance, and are therefore vulnerable to cervical cancer. Men have a prostate and can suffer from prostate cancer. It comes as no surprise, perhaps, that almost all breast cancers are found in women; they have almost exclusive ownership of the specialised cells that make up breast tissue. But there are genuine surprises.

Women and men both have hearts but they get heart attacks in entirely different ways, for instance. They both have livers, too, but a liver disease known as primary biliary cirrhosis occurs primarily in females. The same is true of hepatitis C. Colon cancer in women manifests itself differently from the way it occurs in men. The list goes on, and yet medicine, leaving aside the obvious physiological differences, has treated men and women essentially as the same organism. With the emerging field of gender medicine starting to make a mark, that old way of thinking is starting to pall. As a 2010 *Nature* editorial stated, medicine is finally 'putting gender on the agenda'.

It has taken a long time to get here. One of the earliest reliable papers on sex differences in health came in 1959. That was when two researchers at the University of Würzburg studied the records of nearly 10,000 patients and concluded that men seem to have a lower general resistance to disease. Sometimes, they noted, more than 90 per cent of the people affected by a disease would be of the same sex. A report on this paper in *New Scientist* (*The New Scientist*, as it was then, in its first year of publication) noted that taking into account the sex differences in the prevalence of disease might provide clues to finding novel cures.

The German contribution was quickly forgotten, however. Most reports of the history of this subject start with a 1965 discovery by Thomas Washburn and colleagues at the Johns Hopkins University School of Medicine. They searched case records dating back to 1930 and found 'a significant male preponderance is seen in all the data'. They concluded that the male body was more susceptible to illness.

The beauty of their study was that it covered the era when antibiotics came into use. Having medicines that killed bacteria reduced the overall death rates from infections, but it also skewed the sex ratios: the Gram-negative type of bacterial infection, which was resistant to the first wave of antibiotics, affected more men than women.

The Americans, unlike the Germans, looked for possible causes. Washburn's conclusion was that having two X chromosomes was a good thing when it comes to fighting infections. It is the X chromosome that causes the body to make immunoglobins, or antibodies, that fight disease. Two very slightly different X chromosomes might make two very slightly different types of immunoglobin, the Johns Hopkins researchers reasoned, or simply make more, giving the female body a bigger and potentially more diverse arsenal of weapons with which to combat infection. Those poor males, with just one X (and one Y), would be at a disadvantage.

Clearly they had started something, because the next few years saw a rush of papers noting the sex differences in immune profile. Two years later one group reported that women did indeed have higher levels of immunoglobin M, an antibody manufactured in the spleen, that patrols the circulatory system looking for foreign organisms to disable. Another two years passed, and a group of researchers from London's Royal Free Hospital went one better: they found that the one in every thousand women who has three X chromosomes is even better off. These females, the result of post-conception genetic processes going slightly awry, have even more immunoglobin M than those with only two. Triple X syndrome, as it is known, is associated with a raft of difficulties, many of which manifest early on in difficulties at school, but it does give a boost to the immune system.

The Johns Hopkins study noted that the gender disparity in illness was 'most marked in infancy'. Put simply, the younger the patient the hospital saw, the more likely they were to be male. That's because, when you're young, your immune system is still teaching itself to defend you from attack – girls, apparently, have something of a head start.

It's tempting to say, then, that there's a medical upside to being a girl. But that is true only at the beginning; girls certainly don't come off better in the end. In the 1990s, for example, it emerged that women who had a heart attack aged fifty-five or under were twice as likely to die as men in the same position.

Ask most people – male or female – to fake a heart attack, and they will clutch their chest, complain about their breathing and fall to the floor. This, though, is not how women always experience a myocardial infarction. They clutch their bellies and complain of feeling sick. Lower abdominal pain and nausea are common symptoms of a female heart attack, sometimes supplemented by jaw, neck or back pain. Alarmingly few doctors are aware of this – which is why women complaining of these symptoms don't get an ECG or an angiograph that would be given to a man with chest pain. Even if they do get the right tests, some of the diagnostic tools are less sensitive to detect coronary artery disease in women than they are in men.

If you are a woman over sixty-five years old, you are most likely to die from coronary heart disease. And a woman who has a heart attack and is hospitalised is more likely to die than a man in the same position. Six months after a heart attack, she would be less likely to still be alive than a man of her age. Why? Largely because we have paid a disproportionate amount of attention to the pathology and treatment of male heart problems.

Despite making up more than half the general population, despite being more likely than men to die of cardiovascular disease, participants in cardiovascular drug trials are overwhelmingly male. In 2004, for instance, it was more than three men to one woman. 'Clinical trials on prevention and treatment of cardiovascular diseases have been conducted either exclusively in males or in populations with very low numbers of females,' as Giovannella Baggio puts it. 'Moreover, compared to men, women are less likely to undergo cardiac monitoring, enzyme measurement, recovery in coronary care unit, coronary angiography, and revascularization.'

Baggio has assembled a frightening roster of diseases in which women are at risk because no one has paid much attention to the gender aspect of medicine. Before we get into the specifics of why men and women ail so differently, it makes sense to address this appalling bias. Where did our ignorance come from? From medicine's gold standard – the clinical trial.

In 1993 the US National Institutes of Health admitted it had a drug problem. Women and ethnic minorities were not routinely recruited to participate in drug trials, it said; most studies were reports of what the drugs did to white men. From 1994 the NIH changed its funding rules: women and minorities had to be included in all clinical studies. By 2010, however, the problem still hadn't gone away. In a blistering comment piece in *Nature*, a team of researchers from Northwestern University pointed out that women 'remain underrepresented in biomedical research'. Women were making up only a quarter of subjects in drug trials. Only 13 per cent of studies took gender into account when analysing results. In trials of cardiovascular drugs, women still were not included in proportions that represented the gender ratio of prevalence of cardiovascular problems in the general population.

Because they are severely under-represented in clinical trials of new drugs, women are more likely to have an adverse reaction to a licensed drug than men; for every ten adverse reactions, six will be in women taking the medicine. It seems that only when the medicines are released for prescription and sale does the true clinical testing on women begin. As a result, women's health care has been short-changed: 'medicine as it is currently applied to women is less evidence-based than that being applied to men,' as *Nature* put it. Erika Check Hayden has written even more succinctly: 'The typical patient with chronic pain is a 55-year-old woman – the typical chronic-pain study subject is an 8-week-old male mouse.' That's an interesting observation when we consider Judith Walker's experiments with pain relief.

I'm sure Walker is a lovely person, but she is clearly not averse to inflicting a bit of pain if she has good reason. In 1998, for example, she used electric shocks to discover something shocking about ibuprofen.

Having recruited twenty volunteers, she and her colleague John Carmody tested their subjects' pain thresholds by attaching electrodes to their ear lobes and switching on the power. By ramping up the power and asking the volunteers to rate the pain, the pair established that there was no significant difference in the way they all – male or female – perceived it. Then Walker gave some of them placebos, and the others ibuprofen, and turned the power back on.

This time, things were different. After controlling for placebo effects, Walker and Carmody found that ibuprofen worked to reduce pain for the male volunteers, but not for the females. "This is a paradox," they noted, 'because many of the painful conditions for which nonsteroidal anti-inflammatory drugs are used (e.g., rheumatoid arthritis) occur more often in women.'

The studies are still going on. Though Walker and Carmody are both retired, Carmody has continued trying to get to the bottom of things. Our understanding of the placebo effect has moved a long way forward since 1998, so perhaps, he thought, this is somehow skewing the results. In particular, we know that expectation of pain relief acts as a supplementary analgesic, so Carmody carried out a study that would account for this. In 2012 he published the results in the *European Journal of Pain*. They showed that there is indeed some placebo effect involved. But only for men. Some of the men in the study found that the placebo pills gave them significant pain relief – as did ibuprofen. For the women, it was a different story. As Carmody puts it, 'Our study found that dosages of 800 mg of ibuprofen are ineffective in producing analgesia in women regardless of their expectations.'

It has to be said, the findings have surprised many researchers. In a broad swathe of trials, 400 micrograms of ibuprofen has been found to be effective in both men and women. It may be that this was different kinds of pain. Those studies were for post-operative pain relief, designed to mitigate against the after-effects of dental, orthopaedic, abdominal and gynaecological surgery. Maybe electric shocks are different. However, when it comes to pain, sex differences are not just about ibuprofen.

Anaesthesiologists, for instance, are starting to acknowledge that general anaesthetics affect men and women in different ways, they have different responses to different chemicals. And pity the poor men involved in a drug trial after the removal of wisdom teeth; when they were given a class of painkillers known as kappa-opiods their agony remained at excruciating levels while the women found the kappa-opiods rather effective.

The gender differences in reported pain are striking. Women report higher levels of pain than men when they both have the same condition – whether that is a medical condition or a soft tissue injury. Given that revelation, it is perhaps unsurprising to learn that, when they are freely available, women use more analgesics than men – around 2.4 times as much, even accounting for body weight. You will be surprised to learn, though, that when asked to report the minimum stimulation that causes pain, women demonstrate a lower pain threshold than men. Their pain tolerance – the amount of time for which a painful stimulus can be tolerated – is also lower. That stands against a popular misconception: 66 per cent of women, according to one survey, believe that women are better able to cope with pain than men.

This is not just amusing reportage: it makes a real difference, because women are significantly more likely to receive inadequate pain relief when in hospital. A classic study entitled

'The Girl Who Cried Pain: A Bias Against Women in the Treatment of Pain' found that, when it comes to pain, the system is stacked against females. The authors, Diane Hoffmann and Anita Tarzian, conducted an extensive survey of the literature and found that, controlling for patient weight, nurses give less pain medication to women, and doctors prescribe less. Female patients experiencing chronic pain are more likely to be diagnosed as demonstrating excessive emotionality and attention-seeking behaviour compared to males experiencing chronic pain. That explains why when women complain of pain, they are more likely to be given sedatives where men are given pain medication.

The truth is that the physiological differences between the sexes indicate that women are far worse off when it comes to pain, a discovery that has led several researchers to ask why the difference between the sexes in the way they report pain isn't significantly larger. As Dalhousie University's Anita Unruh put it, 'The question changes from "Why do women and men differ in their experiences of pain?" to "How do women dampen the effect of powerful sex differences in physiological pain mechanisms to achieve only small sex difference in their actual pain experience?"'

Cultural factors don't help. Nurses, for instance, have been conditioned to expect that patients in moderate to severe pain will have elevated vital signs or have their discomfort written all over their faces. Hoffmann and Tarzian make the point that societal training, which teaches women to remain attractive at all times, works against them here. They conclude, 'Medicine's focus on objective factors and its cultural stereotypes of women combine insidiously, leaving women at greater risk for inadequate pain relief and continued suffering.'

Pain is a problematic measure, however. It is subjective, and men and women deal with it in different ways. Brain scans have shown that men and women subjected to the same amount of pain will have different patterns of blood flow in the brain. For the women, there is more going on in the parts of the brain associated with attention and emotion: perceived pain seems to depend on how much attention they are paying to it. Men have their own idiosyncrasies: they will display a much higher pain threshold and report less pain if there is an attractive female in the room. Women do not respond in the same way to the presence of a good-looking man. But the fact is that the gender disparity with pain relief begins with our laboratory studies. Neuroscience experiments use five times as many male animals than female animals. The rumour of female unsuitability for medical trials began way back in 1923, when G. H. Wang monitored the activity levels of female white rats, coupled these observations with a series of vaginal smear tests, and showed that a rat will move around more at certain points in its oestrus cycle.

We know that hormones secreted by the ovaries can certainly change the symptoms a woman experiences when unwell. Anaesthesiologists, for instance, have to adjust their chemical brews based on where a female patient is in her menstrual cycle. But studies indicate that hormone variations don't actually affect the outcomes of experiments as much as everyone who heard about Wang's smear tests imagined. What's more, women – whatever their hormonal state, or indeed whether or not they are pregnant – need medicines that work. At the moment, they are taking some drugs that are, effectively, untested on someone of their chemical make-up.

It's worth remembering, at this point, that men lose out too. It's not easy being a man with osteoporosis, for instance: you don't get picked up by the medical system. 'The proportion of osteo-porotic men undergoing therapy does not reach 1%,' Baggio says. Osteoporosis is considered a largely female disease; even when it should be picked up in men – after a hip fracture, say – there is a problem. Ten per cent of men hospitalised after a hip fracture die; that is twice the figure for women. A year later, mortality is between 30 and 48 per cent for those men, compared with 18 to 25 per cent for the women.

One solution might be fracture-preventing chemicals that we know work for men. However, many osteoporosis drugs have been tested only on women. The bisphosphonate class of drugs is meant to prevent hip fracture, for instance. According to the scientific literature from clinical trials, they bring a 32 per cent reduction in fractures. That might seem like good news for the men who make up more than one-third of age-related fracture victims. The truth is, though, that tests on the bisphosphonate drugs were done on women aged sixty-five to eighty, so we don't actually know what the drugs do for men. For all we know, the biochemical action of the bisphosphonates in the male body could be to reduce bone strength. It's unlikely, because there have been small studies on a few compounds that demonstrated some effects in reducing fractures in males. But they are far from pow-erful studies. It is impossible to know anything for sure without carrying out new, purpose-designed studies, which are expensive and time-consuming.

For all the coronary, cancer and liver disease issues that affect men and women so differently, the most intriguing and heart-breaking gender disparities in medical issues remain those that affect the beginning of life. Generally, females have more vigor-ous immune responses to infectious agents and thus are better at resisting infections. While some deadly diseases don't discrimi-nate, others, such as sepsis, definitely do, killing far more boys than girls. Major birth defects happen in 3.9 per cent of boys and only 2.8 per cent of girls. There are nine major categories of birth defects, and only in the nervous system defects are girls worse off. Boys also seem to be worse off when it comes to communicable diseases, an issue that has been laid at the door of the inferior male immune system.

That said, in most non-communicable diseases, such as asthma and other auto-immune diseases (where the body's immune system harms itself with an over-zealous response), girls fare much, much worse. Nobody really knows why. This is why we have to take up the challenge of gender medicine. But we do have some clues; it seems we might have completely misunderstood how our immune systems work. And to explore that story, we have to go back to 1976, when a young Danish anthropologist was trying to get into a tiny West African country called Guinea-Bissau.

In 1976 Guinea-Bissau had only just won its independence from Portugal. That victory came after a bitter sixteen-year war fought with napalm-carrying fighter planes by one side and the grim determination of an oppressed, impoverished nation by the other. No wonder the victorious Guineans were hesitant to let Peter Aaby, potentially a Western spy, into their newly regained land.

Aaby wanted to spend six weeks studying the tribes of Guinea-Bissau to discover what had spurred each of them to fight the Portuguese. Some tribes had occupied positions of privilege under colonial rule; others had been little more than slave labourers. Somehow the leader of the revolutionaries, Amilcar Cabral, had united them all. Aaby thought the motivations of the tribes would make a fascinating topic of research.

It wasn't to be; Aaby was never licensed to practise

anthropology in Guinea-Bissau. Instead he took the only route into the country that he could find. He came as a lowly medical researcher working with a Swedish charity looking into childhood mortality. Aaby's team was charged with explaining the country's enormously high mortality rate. At the time one in two children died before their sixth birthday. The blame was pinned on malnutrition, and the researchers were charged with finding out just how bad the problem was.

Aaby, setting a standard that he has lived up to ever since, turned the received wisdom on its head. His team examined 1,200 children under the age of six, and found that just two of them were malnourished – and those two were struggling only because their mother had died. There was plenty of food in Guinea-Bissau, the researchers found. It was housing that was scarce.

At the time the average occupancy of a house was three families with eighteen children between them. Any diseases entering the house spread through its occupants like wildfire. During a measles epidemic in 1979 more than one in five of the children infected were killed by the disease – an enormous mortality rate.

The pattern of infection was simple, the researchers discovered. The child who was first infected, the one who brought the infection into the house, usually got out alive. The others fared much worse. They were knocked sideways by the sudden onslaught of a rampant infection that jumps between individuals to create a hugely intense burden on each immune system. There was no relationship between malnutrition and whether you died of measles; it was all about the intensity of the infection. Aaby and his co-workers had disproved a piece of received wisdom. That's why, he says, they were all fired: it was not the answer his employers had been looking for.

Undeterred, Aaby went back to the scientific literature, and to archived records, and found the same phenomenon in data on the 1885 measles outbreak in Sunderland, England. It was repeated in details of epidemics that had occurred in Denmark and Germany. It is now accepted that housing and crowding are the pivotal factors in the devastation wreaked by an epidemic.

Nearly four decades on Aaby is still carrying out medical studies in Guinea-Bissau. Now, though, his focus has shifted slightly. He is still interested in infectious diseases and the havoc they cause. But he is much more focused on why, in Guinea-Bissau and many other developing nations, a disproportionate number of victims are young girls.

Guinea-Bissau was a peaceful country for only a relatively short time. It had roughly two decades of peace after independence. Since 1998 it has been riven by civil wars and military coups and mutinies. Presidents have been shot in the street, and many of the country's historic buildings have been destroyed. It was during the periods of fighting that Aaby noticed something odd.

As a result of the conflicts health centres in Bissau had no Diphtheria Tetanus Pertussis (DTP) vaccine for several months in 2001 and 2002. DTP provides effective protection against the diseases in its name (pertussis is better known as whooping cough). And yet certain children seemed to benefit from *not* getting it.

Those of the right age who were hospitalised during this period were meant to receive DTP and Oral Polio Virus (OPV) as recommended by the WHO. Because of the DTP drought, they received just the OPV. And, shockingly, their case fatality – the number of them who died before their sixth birthday – was three times lower than children receiving both vaccines. Getting DTP, you might conclude, is bad for you. But it's a bit more complicated than that.

As noted, there's nothing wrong with the DTP vaccine per se. However, studies that have been going on for more than two decades show that, when it is administered after other vaccines, such as BCG or measles vaccine, the combination can be deadly. But only if you're a girl.

In the era before routine vaccination, West African girls and boys died in childhood in roughly equal measure. When measles vaccine and the anti-tetanus vaccine BCG were introduced, girls began to outlive boys. When DTP came on the scene, things changed. Among children whose most recent vaccination was DTP, the girls began to die quicker than the boys. This all seems to do with what are called non-specific effects. These are, essentially, unintended consequences of vaccination, and they can be good or bad.

The existence of non-specific effects shouldn't surprise us, really. Vaccination started as something of a wild experiment. Though Edward Jenner is usually cited as the originator of vaccination, he wasn't the first to perform the procedure. In 1774, twenty years before Jenner's work, an English farmer called Benjamin Jesty inoculated his wife and sons. Jesty used a darning needle to take material from a cowpox pustule, then used the needle to scratch his patients' arms. Locals were outraged by this mixing of animal and human material: Jesty was 'hooted at, reviled and pelted whenever he attended markets in the neighbourhood', and some feared his family would start to grow horns. They didn't – and they never caught smallpox, despite repeated exposures.

Even Jenner's work has the mark of dubious science. His first venture was on an eight-year-old farm boy. On 14 May 1796 he scratched James Phipps with a lancet coated in pus from a milkmaid's cowpox sore. On 1 July Jenner then made several punctures and incisions on the trembling and terrified boy's arms, and introduced material from a smallpox patient into them all. Phipps was fine – the experiment was successful, and Jenner wrote a report and sent it to the Royal Society.

As the London Science Museum website points out, had the experiment failed, 'Jenner would probably have kept quiet about it, and James Phipps would have been forgotten, like many other victims of medical experimentation.' Ironically, the Royal Society had no such qualms; they refused to publish the paper because he hadn't done the experiment on enough people and was thus offering only circumstantial evidence.

Hindsight is a wonderful thing, but Jenner's work does now appear to have been worth all the risks – especially when it comes to tuberculosis. The first TB vaccine was used in 1921. Six years later its introduction to a community in the north of Sweden raised the eyebrows of Carl Näslund, the Swedish Tuberculosis Society's physician in charge of the vaccination programme. Näslund compared the survival rates of children who had been vaccinated with the rates of those who hadn't. Ten per cent of unvaccinated children had died, while only just over 3 per cent of vaccinated children had died. It couldn't be accounted for by the protection against TB, because most of the reduction in mortality came in the child's first year of life, but TB generally kills older children. Näslund concluded that the vaccinated children had an improved immune system that protected them against a range of diseases and infections. 'One is tempted to explain this very low mortality among vaccinated children by the idea that BCG vaccine provokes a nonspecific immunity,' he said in a paper published by the Institut Pasteur in Paris in 1932.

Since then controlled trials have confirmed that BCG does something for us on top of protecting against TB. In Britain and the US experiments have shown that BCG reduces deaths caused by things other than accidents or TB by 25 per cent.

In developing countries BCG remains part of the recommended WHO vaccination programme. Here, the beneficial non-specific effects of BCG are clear. The vaccine reduces mortality among newborns by 40 per cent, not by protecting them against TB but by giving them enhanced resistance to otherwise deadly infections by bacteria, viruses, parasites and fungi.

BCG is no longer routinely administered in much of the Western world, where TB is under control (if not exactly in decline), and where infectious diseases are not generally prevalent and lethal. But that doesn't mean western doctors don't use it. BCG may have been developed purely to fight TB, but now it is being taken up as a weapon against cancer. Studies have shown that smallpox and BCG vaccines can reduce susceptibility to lymphoma, leukaemia and asthma. BCG has become the treatment of choice for bladder cancer and has been used to fight multiple sclerosis and type 1 diabetes. Make no mistake about it, those non-specific effects are real – and of real value.

However, what luck gives with one hand, it can take away with the other. While BCG is a boon, DTP, it turns out, could be a disaster for the female immune system.

Your immune system is a complicated beast. It comes in two parts. One is the innate immune system, which is not terribly sophisticated and doesn't distinguish between threats, but it is quick: it responds immediately to anything unfamiliar or untoward, such as the chemicals released when cells are damaged. The other is the adaptive immune system, which arises from the properties of the lymphocyte cells in your blood. These are the roaming police cells that look for suspicious antigens such as bacteria, viruses or cells from another organism. The receptors on the surface of the lymphocyte cells will lock on to features on the surface of the antigen and then cover them in antibody. This disables the antigen and marks it out for destruction by other, specialised killer cells.

The army of killers in the adaptive immune system operates two different signalling battalions, known as type 1 and type 2 helper T-cells. These release chemicals that tell other cells to march into action. BCG and the measles vaccine pump up the type 1 cells, a move that is known to strengthen the immune system's ability to fight off other infections.

A growing body of evidence suggests that unlike live vaccines, DTP tips the balance towards type 2 helper T-cells. Animal studies show that, for unknown reasons, females have a naturally stronger type 2 bias.

This is supported by research carried out by Katie Flanagan, a specialist in infectious diseases at Launceston General Hospital in Tasmania, Australia, who studies the effects of vaccines on the immune response of infants. Her research makes it clear that girls' responses to vaccination differ markedly from boys'. Flanagan thinks it may have to do with sex hormone receptors on the immune cells, but that has not yet been proven. What is clear, she says, is that there are marked differences in the way certain genes get switched on after vaccination. Males hardly switch on any genes, while nearly all the genes are being switched on in females. It is, Flanagan says, 'striking'.

Flanagan's work parallels a study that came out of the University of California, Los Angeles, in 2006. A research team there catalogued the effects of more than 23,000 genes in male and female mice. They found that almost three-quarters of the genes in the fat, liver and muscle tissue produced different amounts of protein, depending upon the sex of the mouse.

Clearly there is a lot more work to do here, but that sex plays a role in physiology and disease is now undeniable. Which means that medicine has a lot of catching up to do. Giovannella Baggio calls it 'a task for the third millennium', one that will require new investment in research, and the 'reorganization of medical teaching and health policy'. It's no small task – rather like the subject of our next chapter. However, it seems that we might finally be ready to take a project that Sigmund Freud abandoned in the nineteenth century and turn it into a twenty-first-century revolution. Psychobabble has finally found its scientific feet.

6

WILL TO LIVE

Your mind has power in your body

Psychotherapy is a biological treatment.

Eric Kandel

I n 1985 William Keatinge, a physiologist based in London, carried out a remarkable experiment. He and his colleagues persuaded a man to get into a treadmill pool of neck-deep water wearing a shirt, sweater and jeans. The water was bitterly cold – just above 5 °C – and the man ran the risk of hypothermia. The researchers instructed him to move his arms and legs to mimic swimming for as long as he could stand the cold. Survival time should have been around an hour and fifteen minutes. Ten minutes after the survival time had been reached, the researchers halted the experiment. Not because the man was cold, but because he said his feet hurt.

The man in question was Gudlaugur Fridthorsson. It is astonishing that he agreed to take part because the experiment was a re-enactment of the moment, seventeen months earlier, when he almost lost his life.

Fridthorsson is an Icelandic fisherman. On 11 March 1984

stormy seas rolled his boat onto its side 5 kilometres east of the Westmann Islands, an archipelago off the southern coast of Iceland. The vessel turned upside down, and for three-quarters of an hour, Fridthorsson and his two remaining colleagues – another two never made it out – sat on top of it.

Then the boat sank. The temperature of the sea that night was 5 °C. The air above it was at –2 °C. The trio had no choice but to swim for land. They could see the faint spot of a lighthouse, and agreed to head that way, and to keep calling to each other as they swam. Within a couple of minutes, one of them had stopped calling. Fridthorsson and his captain, Hjortur Jonsson, kept swimming and talking. Less than ten minutes later, Jonsson had stopped replying, and Fridthorsson knew he was now alone in the vast ocean.

For six hours he swam in jeans, a shirt and a jumper. He talked to the fulmars wheeling overhead. He talked to God. He told himself jokes. And then, in the middle of a dark night, he reached the shore.

How did he survive? Partly, by being fat. In their paper in the *British Medical Journal*, Keatinge and his colleagues talked about Fridthorsson's 'greater fat thickness' than many subjects in comparative studies. *New Scientist* was not so polite. 'With the greatest respect to Fridthorsson, it is impossible not to be reminded of a seal and its sheath of blubber,' is how writer Stephen Young put it. But there was also the positive attitude: the joke-telling and the conversation with the Almighty.

According to the US National Oceanographic and Atmospheric Administration (NOAA), the average damaged ship takes at least fifteen to thirty minutes to sink. That half-hour 'provides valuable preparation time' in which to get yourself ready for entering the freezing water. The NOAA advice is to wear several layers of clothing: even wet clothes will trap some warm air. Get to a lifeboat if you can, obviously. If you are able to control your entry into the water, lower yourself in slowly to avoid shocking your cardiovascular system and speeding up your breathing. Keep still – don't attempt to swim. Just float, unless there's a lifeboat nearby. NOAA's final piece of advice to anyone trying to survive immersion in freezing cold water is this: 'Keep a positive attitude. Will to live makes a difference.'

Does it? That is the question we are about to explore. Here is an interesting place to start: in July 2012 Helena Karppinen published an extraordinary study called 'Will-to-live and Survival in a 10-year Follow-up Among Older People'. It looked at four hundred inhabitants of Helsinki, all aged between seventy-five and ninety. They had first been interviewed in 2000, when they were asked the innocuous question 'How many years would you still wish to live?'

Some people didn't answer the question, as if they were afraid of putting their thoughts into words. The statistics of this group are interesting: they were older than those who did respond, and had more health problems. Death, to them, was too close to risk conjuring into reality with words.

Karppinen divided the subjects who did respond to the question into three groups, the ones that wished to live for five years or fewer (26 per cent of them), those who expressed a desire to live between five and ten years (56 per cent) and those who said they wanted to live for more than ten years (18 per cent).

By the time Karppinen came back to them ten years later, half of the subjects had died. She and her colleagues looked at how the deaths were distributed, accounting for age, gender, education, smoking, depression and symptoms of illness, and came to a remarkable conclusion. 'Those who wanted to live longer also survived longer,' she said.

You are probably not terribly surprised by this. Human

experience tells us that having the will to live does indeed make a difference. Science, ever cautious about the way we interpret the world around us, has fought tooth and nail to deny it. The intensity of that fight, though, seems to be abating. The field of psychoneuroimmunology, the study of the interplay between the brain, the mind and the body's defences, is finally coming in from the cold. Which is appropriate, since the whole thing started in Russia.

The story of what has come to be known as psychoneuro-immunology (PNI) began on 15 November 1885. That was when a rabid setter called Pluto sank his teeth into the flesh of Alexander Demiankenkov, an officer serving in the St Petersburg Corps of Guards. Demiankenkov's commanding officer was Prince Oldenburgskii, a member of the Russian royal family and a keen student of medical research.

Prince Oldenburgskii had heard about a remarkable cure for rabies created by Paris's celebrated medical scientist Louis Pasteur. Pasteurisation – killing bacteria by heating – was only one of the Frenchman's discoveries. His creation of a suite of vaccines was just as significant, and the rabies vaccine, the toast of a rabies-infested continent, had just been developed. The prince sent his injured officer to Paris for treatment. The following year Russia made a formal request that Pasteur train their scientists in his techniques. Pasteur wasn't going to give away his secrets cheaply. He suggested it would be better that the Russians 'provide financial aid' to his Paris institute. On receipt of such funds, Pasteur declared, he would consider himself 'fortunate to receive in this institute physicians from your expansive country, who always elicit in me great sympathy'. In the meantime, he advised, the Russians should send all bitten personnel to Paris with due haste.

The tactic worked. Russians soon became the largest contingent of patients at Pasteur's Institute, and Prince Oldenburgskii was dispatched to France with a gift of 100,000 francs from Tsar Alexander III. He did not return empty-handed: Pasteur gave him a rabbit that had been inoculated against rabies – the perfect starting point for the development of a Russian anti-rabies vaccine.

Pasteur also stood by his invitation to Russian researchers to come and work in Paris. The Institute's research papers, he said, should be published in the languages of both nations. And that was how Sergei Metalnikoff came to Paris. He brought with him the conditioning techniques of Ivan Pavlov, which he used to demonstrate that he could affect the white blood cell count of a guinea pig by messing with its mind.

White blood cells make up about 1 per cent of your body's blood. Their role is simple: to attack foreign matter found in the blood and render it harmless. A raised white blood cell count suggests that your body's defences have been activated. And, as Metalnikoff proved, sometimes that's possible even when there's no enemy in sight.

Metalnikoff's experiments involved placing a warm bar on a guinea pig's skin at the same time as his collaborator Victor Chorine injected the animal with tapioca. The tapioca triggered the animal's immune system, and increased the white blood cell count. Following Pavlov's conditioning protocols, however, Metalnikoff soon made the injection unnecessary. After a few rounds of conditioning, the sensation of the warm bar alone was enough to raise the animal's white blood cell count. As it turned out, the process was advantageous to the guinea pig. By the end of their experiments, the application of a warmed bar to their skin made the conditioned animals far more resistant to a cholera infection than their unconditioned cage mates.

Metalnikoff's results were published in 1926 but no one took much notice. That's because the empire of the mind now

belonged to psychoanalysts and psychiatrists, and no serious scientist would work with such charlatans. It has to be said, Sigmund
Freud is largely to blame.

In 1936 Princess Marie Bonaparte, great-grandniece of Emperor
Napoleon I of France, was browsing in a Berlin bookshop when
she discovered a sheaf of letters Freud had written to his friend
Wilhelm Fliess. The princess knew Freud well. She was his wealth-
iest acolyte, and had paid for his passage out of Nazi-occupied
Austria. She bought the letters, then wrote to Freud, asking what
she should do with them. Burn them, he said.

Perhaps she was aware he wasn't in the best frame of mind
to make such a decision. Too many cigars had left him riddled
with a cancer of the jaw that had collapsed the right side of his
face. After numerous operations he was in intolerable pain; Vir-
ginia Woolf described him as 'a screwed up shrunk very old man:
with a monkey's light eyes'. For whatever reason, the princess dis-
obeyed Freud's command, and we should be grateful because the
letters, finally published in 1950, are illuminating.

The letters contain a description of Freud's 1895 attempt at
a 'Project for a Scientific Psychology'. No wonder Freud was
embarrassed: the project was a miserable failure. He was a trained
neurologist, and his idea was to find the neural basis for human
behaviour. Brilliant though he was to see the necessity of the
project, the contemporary understanding of the brain's physiology was poor – certainly nowhere near advanced enough to
shed light on the quirks of humanity being exposed in Freud's
consulting room. Realising it was a waste of time, Freud abandoned the half-formed project and declared that psychoanalysts
should base their diagnoses only around the subjective reports of
patients. Attempts at linking psychoanalysis to anything scientific

were worthless, Freud said. Psychoanalysis and objective science parted company almost immediately afterwards, and only now are they experiencing even the beginnings of a rapprochement.

The reconciliation started, appropriately enough, with a psychiatrist's father–son relationship. In the 1950s George Solomon held a residency in psychiatry at the Langley Porter Institute of the University of California, San Francisco (UCSF). His father, Joseph, also a psychiatrist, had long suspected that psychological issues affected the course of rheumatoid arthritis; the condition was exacerbated by negative emotions, or stressful circumstances, he felt. George decided to help his father and investigate whether this was true.

Eventually George Solomon moved from UCSF to Stanford University. There he began a collaboration with psychologist Rudolf Moos, who was investigating the range of factors that influence the health of patients with rheumatoid arthritis. The effects of the mind, however, could only be studied with patients back in San Francisco; the Stanford authorities were having none of the pair's 'psychological nonsense'.

By the 1960s the UCSF department wasn't quite so aghast at the idea of a mind–body connection. In the previous decade some of its researchers had managed to show that stress conditioning could make mice more vulnerable to the herpes virus, and that stressed monkeys were more likely to succumb to polio. Nonetheless, it was quite a shock when Solomon nailed his colours to the mast. He hung a sign on the door of his laboratory. It said, to the horror of his colleagues, 'Psychoimmunology'. A new field of research was officially born.

The immunologists were particularly nonplussed. The human immune system had no need of the brain, they declared: it was perfectly able to recognise for itself the various alien molecules and organisms that posed a threat to the body. What's more, there was no known physical connection between the brain and the immune system. Luckily, no one told Robert Ader this.

'As a psychologist, I was unaware that there were no connec-tions between the brain and the immune system,' Ader admitted years after his experiments had proved received wisdom wrong. He had been studying how rats learn by conditioning, pairing saccharine drinks with injections of a drug called cyclophosphamide, which gives the rats a stomach upset. Unsurprisingly, the rats learned to avoid the sweet taste of saccharin. During the course of the experiments, however, rats began to die, an outcome that Ader initially saw as a 'troublesome but uninteresting observation'.

Soon, though, it became interesting. He noted that the deaths were related to the volume of saccharin consumed rather than the amount of cyclophosphamide injected. Not being an immunologist, Ader was free to make wild, uninformed speculations. He knew cyclophosphamide suppressed the immune system; perhaps a conditioning-induced link was enough to suppress the immune system too? With a compromised immune response, the rats would be more susceptible to any pathogens around the lab – hence the deaths.

Entirely unaware of what Metalnikoff had done with tapioca and guinea pigs half a century before, Ader and his collaborators went on to prove a relationship between conditioning and the immune system, forcing the world to confront a hitherto undiscovered connection between the immune system and the brain. 'I was "forced" into it by my data,' is how he put it later. And thus began the farce that followed Schopenhauer's Law so beautifully.

The human will was philosopher Arthur Schopenhauer's favourite subject. He discussed the 'will to power' and the human spirit embodied in a 'will to live'. He was aware that stubbornness and inability to lose face was also part of the human condition, an observation summed up in one of his more insightful statements. 'All truth passes through three stages,' he said. 'First it is ridiculed. Second it is violently opposed. Third it is accepted as being self-evident.'

This is exactly what Ader and his colleagues encountered. We don't yet know that the idea of a physiological will to live is what Schopenhauer calls 'truth'. But Ader had papers showing enhanced antibody production through psychological conditioning rejected by the British journal *Nature* because he was unable to give the details through which such enhancement might occur. A few years later, in the late 1970s, *Nature* rejected a paper showing that the nervous system can respond to signals from the immune system because 'it is self evident that the brain must receive information from the immune system.'

Another measure of how far we have come can be found in the fact that, in 2000, an Austrian-born psychiatrist won the Nobel Prize for work showing that the mind is a physical entity.

When Eric Kandel won his Nobel Prize, Austria was ecstatic – but not for long. Kandel says he 'stuck it to the Austrians' by declaring publicly that it was not an Austrian Nobel, but a Jewish-American Nobel.

Kandel is not well-disposed towards the Viennese people who, he says, have never faced up to their exile of the Jewish population. In his autobiography, *In Search of Memory*, he asked, 'How could a highly educated and cultured society, a society that at one historical moment nourished the music of Haydn, Mozart, and Beethoven, in the next historical moment sink into barbarism?'

Shortly after this public excoriation of the country of his

birth, the Austrian president phoned Kandel and asked what the country could do to make things right. Kandel's first demand was that Doktor-Karl-Lueger-Ring, the road on which the University of Vienna is situated, must be renamed.

Karl Lueger was elected mayor of Vienna just before the turn of the twentieth century. He was a strident Catholic, and viciously anti-Semitic; Adolf Hitler, who was a resident of Vienna for six years during Lueger's mayoralty, cited him as an influence, and Lueger even won a citation in Hitler's *Mein Kampf*. Hitler's eugenics came from the distortion of Darwinism via views such as Lueger's, and Kandel describes it as 'an incredible insult' that Lueger's name should still be on the city's landscape, particularly in close association with the university that had nurtured such stellar Jewish thinkers as Boltzmann and Freud. In April 2012 the city renamed the street Universitätsring. Kandel is a man who gets things done. And that includes changing our minds about the nature of the mind.

The physicality of memory – its effects on the physical structures of the brain – has played a huge part in Kandel's life. He still remembers fleeing Vienna in the months before the Second World War broke out. His family's apartment at Severingasse 8, in Vienna's 9th District, had been ransacked by anti-semitic thugs. The Kandels were forced out into the street, and all their possessions stolen. The Nazis robbed the nine-year-old Kandel of his most prized possession, a blue remote-controlled toy car; the memory of it was indelibly imprinted on his brain. 'I am struck, as others have been, at how deeply these traumatic events of my childhood became burned into memory,' he says. 'I cannot help but think that the experiences of my last year in Vienna helped to determine my later interests in the mind, in how people behave, the unpredictability of motivation, and the persistence of memory.' He won his Nobel Prize for showing that the storage of memories is a 'very molecular process', as Kandel put it in his banquet speech before the king and queen of Sweden and the members of the Nobel Assembly. And it all started with a snail.

Animals have served us well when it comes to understanding the hidden communications systems of biology. We came to understand electrochemical action potential, the way neurons send signals to other cells, through studies of the squid's anatomy. Its giant axon, a current-carrying biological wire that stretches from the squid's head to its tail, is a millimetre in diameter, far broader than anything in our brains and perfect for scientific study.

We also owe a debt to the horseshoe crab, whose huge compound eyes and easy-to-observe optic nerves made it so easy to understand the interplay of the retina and the brain. As every student of biology knows, we have the frog to thank for our grasp of the interactions between nerve and muscle cells.

Nonetheless, everyone told Kandel that researching a snail's memory was a fool's errand. No one knew anything about how memories might be held; while elucidating the precise mechanisms of memory struck Kandel as a 'wonderful' problem to solve, his seniors told him that the brains of invertebrates – creatures without a backbone – were too simple to act as a model for human brains. But Kandel knew that even some of the simplest animals could learn. Therefore, he reasoned, some of the primitive structures of their brains might be preserved within our own physiology.

Scanning through a list of candidate animals, rejecting crayfish, lobsters and nematode worms, he eventually decided that the giant marine snail *Aplysia* would fit the bill, and that he would learn everything he could about them from the world's expert,

Ladislav Tauc. And so it was that Eric Kandel went to study snails in Paris.

Aplysia is so useful precisely because its nervous system is so simple. It has relatively few cells, all of which are large and individually identifiable. In other words, it is easy to follow what each cell is doing.

The first step was to show that *Aplysia* can learn. Kandel and his colleagues, watched by highly skeptical colleagues, managed to condition the animals' behaviour so that they would retract their gills in a defensive manoeuvre when given a light tap.

Normally the animals wouldn't respond to a touch. But combine a few touches with a nasty electric shock, and they soon learn to fear the researcher's finger. The burning question was, with the snails' simple anatomy, could there be a visible difference between the brain of a trained and an untrained snail?

There was. When a snail has remembered something for the long term, examination of its brain reveals that it has grown new connections, or synapses. Kandel was simultaneously excited and completely unsurprised by this discovery. A trained psychiatrist, he only began to explore memory because he was looking for a scientific breakthrough that would inform and improve his field. He has no doubt that his discovery with snails ought to be true of humans too: human learning must involve changes to the physical brain, he reckoned. What's more, he believed that psychotherapy was essentially a learning process, and so psychotherapy must change the brain. Here we can finally connect the state of the physical human body and the contents of its mind, he said. It's quite a claim, coming as it does via research on a snail, but we now know, thanks to a brain imaging technique called Positron Emission Tomography (PET), that it is entirely justified.

PET scans have revolutionised medicine, both in diagnosing patients and in research. All it requires to start things off is an injection of radioactive sugar. The radioactive element gives off positrons, the antimatter equivalent of an atom's electrons. When these positrons meet electrons in your tissues the result is annihilation, with a flash of energy. That flash, more accurately described as gamma radiation, is picked up by the scanner. By noting where the radiation hits the detectors in the scanner, computer software pieces together its trajectory and, from that, the place where the annihilation took place. The result, thanks to some ingenious engineering, is a three-dimensional map of your body.

It can be an incredibly high-resolution map. Different kinds of tissue take up different radioactive elements in differing amounts, so an initial injection of a cocktail of elements allows researchers to create three-dimensional models of your body tissue at a molecular scale. They are astonishingly revealing.

At the University of Eastern Finland, Soili Lehto assembled nineteen depressive patients and put them through a PET scanner before and after a year's psychotherapy. The scans measured the density of the serotonin transporter molecules in the central region of the patient's brain. Most of the patients showed no change in the density of the serotonin transporters. But eight of the patients showed a significantly increased transporter density after psychotherapy.

Those eight were identified as 'atypical' depressives, which means their mood could be lifted on occasions when good things happened. With such a small sample, and no control group, it's difficult to know what the results mean. But they did give definitive evidence that psychotherapy can change the brain. We got more evidence in 1992, when California neuropsychiatrist Lewis Baxter compared psychotherapy with the effect of the antidepressant fluoxetine (better known as Prozac). Both treatments produced changes in the activity of the caudate nucleus of the brain, a structure that sits under the cerebral cortex.

While we expect that chemicals have a physical effect on the brain, the idea that talking has the same effect seems surprising, to say the least. Nonetheless, it's clear that it does. In 2011 Hasse Karlsson, professor of psychiatry at the University of Helsinki looked at twenty studies of brain changes induced by psychotherapy and concluded that we are moving towards a situation where we know so much about what psychotherapy does – how our subjective experience can be manipulated to change the physical structures of the brain – that specific types of psychotherapy can be used to target particular brain circuits. As Kandel puts it, 'psychotherapy is a biological treatment, a brain therapy.'

The programme that Freud abandoned is finally bearing fruit. So how far can it go? We now know that the mind can affect the immune system, and that talking through problems can cause physical changes to our brains. We know that certain states of mind – expectation of an assault, for example – can alter the immune system, making organisms more resistant to infection. So, can we say, with NOAA, that having the will to live makes a difference? Have we hit upon the holy grail of healthcare?

Not quite yet. Before we go too far, it's worth revisiting the question of whether there really is an effect worth harnessing. The singular case of Gudlaugur Fridthorsson leaves us intrigued but unsatisfied – he was, after all, protected by his fat. Helena Karppinen's study of elderly people wishing themselves a longer life is a little more convincing. But what other reasons do we have to take the physiology of the will to live seriously? Perhaps we should start with the extraordinary claims of David Phillips.

Phillips, a sociology professor at the University of California, San Diego, is no stranger to trouble. He was born in South Africa, where his Lithuanian Jewish parents expressed their outrage at apartheid by delivering healthcare exclusively to the non-white population. In the 1950s, when David was twelve years old, the family was forced to flee to the United States. By 1974 he was a grown-up dealing with death: that was when he published a hugely influential paper on 'copycat suicides'. The paper showed that publicising a suicide – putting it on the front page of the newspaper, say – dramatically increases the number of suicides occurring in the following few weeks. Though hugely controversial at the time, Phillips's finding has stood up to scrutiny, and changed the way the media now reports on people taking their own lives.

Also controversial was his article, published in the *New England Journal of Medicine*, that claimed to show how scientists are influenced by research discussed in the *New York Times*. Publish a paper in the *New England Journal of Medicine*, then get it featured in the *New York Times*, and it will get far more citations from other scientists than an equivalent article that the *New York Times* ignored. Scientists hated the finding, and worked hard to find the flaw in Phillips's analysis. They failed to find one.

Nothing Phillips has published, though, has been as controversial as his studies on people delaying their own deaths. He started by asking an interesting question. 'In the movies and in certain kinds of romantic literature,' he said, 'we sometimes come across a deathbed scene in which a dying person holds onto life until some special event has occurred. For example, a mother might stave off death until her long-absent son returns from the wars. Do such feats of will occur in real life as well as in fiction?'

To answer it, Phillips first examined the birthdates and

deathdates of 1,251 'notable' Americans. It would be reasonable, he said, to assume that famous people's birthdays will be celebrated with more pomp and ceremony, that people will lavish more attention and gifts on them, and that their birthdays will be, generally speaking, much more worth hanging around for than the birthday of some ordinary schmuck. Thus there should be a 'death dip'; the famous should be less likely to die just before their birthday than just after.

Phillips did see such an effect. He also found a correlation between the degree of fame and the significance of the death dip. He grouped the notables into three tiers of fame, with the likes of George Washington and Thomas Edison in the top group, Edgar Allen Poe and Alexander Graham Bell in the middle and Samuel Adams and Nikola Tesla among the least notable in the study. The more famous the people in the group, the bigger its death dip. If you are a big cheese, you are unlikely to die just before your birthday, Phillips declared – largely because you crave attention.

The 'Deathday and Birthday: An Unexpected Connection' study, published in 1972, has its fair share of detractors and debunkers. The same can be said of Phillips's follow-up 'Death Takes a Holiday: Mortality Surrounding Major Social Occasions'. This was published in the Lancet in 1988, and appeared to show that a death dip occurs among Jewish males in the week before Passover. This celebration is led by the male head of the household, and the dip was deepest when the holiday fell on a weekend – presumably because the celebrations would be even bigger, and definitely not to be missed.

Are there examples for other cultural groups? Indeed there are: Phillips found that the Harvest Moon Festival, where elderly Chinese women are the centre of attention, is preceded by a death dip among this very group. The week before the festival, the matriarchs were one-third less likely to die than normal. The week after, they were more likely to die than in an average week. The variation each side of the holiday was roughly the same. Again the data pointed to the possibility that we can will ourselves to stay alive – for a short while, at least.

Plenty of researchers have accused Phillips of cherry-picking his data to fit the hypothesis. But looking forward to something – and optimism generally – does seem to have positive effects on your health. In a 2001 study Harvard University's Laura Kubzansky found that optimism – 'viewing the glass as half full', as she puts it – reduces older men's susceptibility to coronary heart disease. The finding was independent of how much they smoked, or how much alcohol they habitually drank. It's possible, Kubzansky said, that training people to be more optimistic would be a worthwhile investment in healthcare.

Before we delve too far into this area, it's important to note the risk of 'saccharine terrorism'. Telling those who are ill that if they only think positive thoughts they'll get better can have disastrous effects; those who don't get better can be made to feel it's because they aren't being upbeat enough. They then take on unnecessary responsibility for their continued illness. Neither case is true. 'In their enthusiasm to advance positive psychology, its advocates have created an enormous gap between their assertions and scientific evidence.' So say James Coyne and Howard Tennen in an article written to burst the bubble of the exaggerated claims of psychological effects on cancer. They quote a 1999 study of 578 women with breast cancer where the myth of the power of positive thinking was busted wide open to the relief of tired, guilt-ridden patients everywhere. 'Our findings suggest that women can be relieved of the burden of guilt that occurs when they find it difficult to maintain a fighting spirit.'

That said, it's also important to be aware of the destructive power that adversity brings to your body.

Stress is a many-headed monster, but it essentially describes what happens in your body when you feel ill-equipped to deal with the demands you are facing, or that you have lost control of what is happening to you. The fight-or-flight adrenaline rush is a good example of your response to a certain kind of stress; being mugged, for example, will send a chemical cocktail to various of your muscles, equipping you to escape or attack your assailant.

Whether it's a mugging, taking an exam, parachuting or burnout at work, a stressful event causes the central nervous system, the endocrine system and the immune system to exchange a host of chemicals: adrenaline, noradrenaline, cortisol and pro-lactin to name but a few. Every one of them can adversely affect the immune system because almost all of our immune system cells have receptors that will bind to the stress chemicals. The result is a release of cytokines.

Cytokines are the killer proteins of your immune system. They go by names such as interferon, Tumour Necrosis Factor (TNF) and interleukin; released by white blood cells, their job is to attack, neutralise and destroy invading organisms or any molecules that pose a threat to the body's normal functioning. Sometimes, though, the cytokines are unleashed because of stress. That's a problem because the cytokines have their own effect on the chemical soup inside your body. Interleukin-1, for example, can cause the brain's hypothalamus to release a hormone that stimulates the production of more stress hormones, escalating the whole situation. Studies have shown that elevated levels of stress hormones and cytokines make you more susceptible to colds and influenza viruses, less responsive to vaccines and slower to heal when your skin is cut or subjected to a lesion. In slightly bizarre experiments where women caring for a relative with Alzheimer's disease (a high-stress situation) allowed themselves to be cut with a 'small, standardised dermal wound', the wound took 24 per cent longer to heal than it did in women in the unstressed control group. As if to add insult to injury, the researchers took blood samples from the women. What they found was that the leukocytes in the stressed carers' blood produced less infection-fighting interleukin-1-beta when put in contact with an antigen.

There is little in life that is more stressful than the death of a loved one – and here the destructive power of stress is clear. A report written for the National Institutes of Health in 1994 said that, 'in study after study, the mortality of the surviving spouse during the first year of bereavement has been found to be 2 to 12 times that of married people the same age.' It's known as the 'widowhood effect'. Another piece of research, carried out on 4,000 people over a twelve-year period, showed that a man whose wife has just died had a 25 per cent higher chance of dying in those twelve years.

It's worth pointing out that the increase in mortality for a woman whose husband has died is just 5 per cent. That is almost certainly due to the relative self-sufficiency of the women and men in this study; they are of a generation where widowed men tend to die from the lack of care and cuisine as well as lack of companionship. Researchers are well aware of this. A meta-analysis – essentially, a study of studies – published in 2011 laid out a number of factors that could be responsible. Having a partner allows you to share costs, making it easier to make healthy food choices. Spouses monitor each other's health, for instance, and make sure the other follows healthcare advice and complies with medical regimens – the 'guardian effect'.

Nonetheless bereavement is certainly a problem, whether you are male or female. Between 1975 and 1977 Roger Bartrop and his colleagues at the University of Sydney measured the immune system responses of twenty-six recently bereaved widows and widowers. In the first eight weeks after bereavement, their immune system was far less responsive, putting them at a greater risk of acquiring infections.

That was the first time severe psychological stress had been shown to produce a measurable difference in the human immune system. A follow-up study published in 2010 has shown that it is not just infection that you should fear if bereaved. Most types of illness were only 20 per cent more prevalent in the bereaved group than the control group. But all of the bereaved reported heart and circulatory problems twice as often as the people in the control group. Stripped of the optimism that Laura Kubzansky has shown is so protective of the cardiovascular system, their hearts were literally broken.

Remaining unmarried is no protection. In 2010 a study enumerated the dangers of loneliness. if you have 'adequate' social connections you are 50 per cent more likely to live to the end of a specified period than those who are lonely. In other words the effect of good friends is roughly similar to giving up smoking or making a significant cut to your intake of alcohol. A 2012 study, which followed 2,000 US citizens aged fifty and above, found that being chronically lonely was associated with being almost twice as likely to die over the period of the study.

The pain of recent separation is almost certainly worse than the sting of ongoing loneliness. The widowhood effect is 'markedly stronger in the months immediately following the spouse's death than in the years following,' according to the 2011 meta-analysis. It's the six months after bereavement that will kill you. Survive those, and your risk of dying moves back to near-normal. Why? They suggest that the most plausible explanation involves the psychoneuroimmune pathways. It's time, as we come towards the conclusion of this chapter, to delve a bit deeper into the immune system.

When you are ill, you exhibit certain behaviours. That feeling of fatigue, the loss of appetite, the fever and the chills and the aches are not what a virus or bacterium do to you. They are what your body's defence mechanism does to you. All those symptoms are the results of cytokines: they cause 'sickness behaviours', which make sure they have enough resources to do their job. So they knock you off your feet so that they can get on with the job in hand, secure in the knowledge that you're not going to use up all the body's energy running a marathon that day. 'Sickness is a normal response to infection, just as fear is normal in the face of a predator', says Robert Dantzer in a 2008 paper.

Sometimes, however, the sickness response gets out of control – and your mind can be the innocent casualty. Twenty years ago, clinicians trying to boost the immune system of people whose tumours had resisted radiotherapy and chemotherapies, or whose hepatitis C wouldn't budge, injected them with cytokines. It produced all the expected sickness behaviours. It also, in many cases, produced depression. After that discovery, researchers began to examine the blood of depressed patients, looking for markers of an overactive immune system. They found them, but no one dealing with depression wanted to know. The novelty of this discovery 'failed to attract the interest of the psychiatry community', as Dantzer puts it.

Around a decade later that changed. We now know that one-third of people diagnosed with major depressive disorder will have raised levels of inflammatory markers like cytokines in their blood. Disruption to the immune and inflammatory system has been found in patients suffering post traumatic stress disorder, and with schizophrenia. Treat certain depressed patients with the frontline drugs and their blood markers for inflammation are significantly reduced.

We are finally appreciating that what goes on in the mind is not independent of what is going on in the body. As Glaswegian researchers Rajeev Krishnadas and Jonathan Cavanagh put it in a paper published in 2012, 'we now know that the brain is not an immune privileged organ.' Changes in the inflammatory system can change the way the brain works. What's more, changing the brain can change the body's functioning.

For the most part, that comes down to reducing stress where possible. A 2003 study published in the *Journal of Psychosomatic Medicine* showed that the practice of T'ai chi – meditation through movement – can boost cellular immune responses by up to 50 per cent. When David Spiegel of the Stanford University School of Medicine began to investigate the effects of support groups for people with cancer in the 1970s, he and his team were concerned that watching others die of the same disease would demoralise patients and perhaps even hasten their demise. It didn't, though: putting them in a room to talk to each other for an hour once a week, a process he calls 'supportive-expressive group therapy', extended their lives, according to research he published in 1989. They lived eighteen months longer than the control group patients who had equivalent cancer burdens but didn't participate in organised sessions.

At the time plenty of people were skeptical, but the skepticism is draining away. In a 2012 study, Spiegel reports eight studies that have corroborated the evidence that psychotherapy or other 'psychosocial' interventions have improved survival times. Over the same time period six similar studies showed no such benefit. The worst you can say, Spiegel points out, is that there was no harm done. The best is that people lived longer, with a better quality of life because they took measures to reduce their stress.

Even if there isn't a will to live, psychoneuroimmunology is telling us that there is at least a will to live longer. Eric Kandel has outlined where he felt psychoneuroimmunology should go. As our understanding of the brain grows, so will our ability to use it to take more control of our health. What we call mind, he says, 'is a set of processes carried out by the brain.' It is, he said, 'likely that in the decades ahead we will see a new level of cooperation between neurology and psychiatry.' The result will be a great leap forward for human well-being.

It's not clear we can say the same of our next topic. The role of quantum physics in biology is only just coming to light, and there's no indication we will ever use such insights to impact our health. It could still change the face of technology and save the planet from catastrophic climate change, however – and that makes it well worth a look.

7

CORRELATIONS IN CREATION

Biology is putting quantum
weirdness to work

*In nature hybrid species are usually sterile, but in science the
reverse is often true.*

Francis Crick

There is something wonderfully ridiculous about a robin
wearing an eye patch, but there doesn't seem to be any lasting
shame. When Roswitha and Wolfgang Wiltschko let the birds go
from their laboratory at the Goethe University in Frankfurt, they
always flew freely away.

These birds are the lucky ones – and not just because they get
to look like a pirate. The Wiltschkos let their robins go after the
winter is over, when they are itching to return north to Scandina-
via. Months earlier the hundreds of millions of birds that didn't
get trapped by the Wiltschkos headed south past Frankfurt and
on towards north Africa. Every year, as they pass over the southern Mediterranean, around 15 million of them are illegally caught
and killed for food. Cyprus is the worst culprit: songbirds are a

local delicacy on the island, cruelly trapped to suffer a long, lingering death before becoming a 40-Euro platter of *ambelopoulia*.

That's not to say that the birds wintering with the Wiltschkos in Frankfurt are entirely without stress. After all, they have to withstand the indignity of an Emlen funnel, a large flat-bottomed bowl with a lid on top. The sides are lined with paper, and the bird stands on an ink pad placed in the bottom of the bowl. Then you do something mean, such as putting the bowl at the centre of a planetarium. When the bird sees the stars overhead, it will try to fly in a certain direction. As it does, it slips down the side of the bowl, its feet leaving tell-tale ink-marks on the paper that show the direction in which it tried to head.

You don't have to have a planetarium to confuse the birds, though. The Wiltschkos regularly used an artificial magnetic field, for example. With a robin in the bowl, and the bowl in a box that screened out the Earth's magnetic field, Wolfgang Wiltschko would put an electric current through large coils of wire to create his own field. He could orientate it whichever way he chose. The ink marks on the paper showed that the robins were fooled: in the autumn they would try to fly in whatever direction Wiltschko had made seem like slightly west of south. If he did the experiment in the spring, they would fly slightly east of north. Why? The answer would seem to be that they are in possession of quantum superpowers.

Roswitha and Wolfgang Wiltschkos' strange work is a continu-ation of something started half a century ago, when the physicist Erwin Schrödinger published a book called *What is Life?* Though only a slim volume, it had a huge impact, and inspired a generation of physicists – including the co-discoverer of the structure of DNA Francis Crick – to take up biology.

Like everyone who has tried since, Schrödinger failed to come up with a decent answer to his question. But he did have an intriguing take on the problem. The first question to answer, he said, is about the difference in scale between things that are quantum and things that are alive. As Schrödinger put it, 'Why must our bodies be so large compared with the atom?'

The answer to that question comes from an examination of the central currency of quantum theory: uncertainty. Quantum theory abounds with probabilities, possibilities, maybes and things that are neither here nor there, all the while being simultaneously here *and* there. In 1943, at the time when Schrödinger was giving the lectures in Dublin that were turned into *What Is Life?*, physicists understood quantum theory through the application of statistics. You couldn't predict the outcome of a measurement on a single quantum particle, but if you measured enough of them, you began to see a pattern in the results. This was what Einstein famously rejected with his declaration that god would never 'play dice' with the universe.

The thing to note, Schrödinger said, is that despite the uncertainty, life is stable. We reproduce and – occasional, exceptional circumstances aside – the offspring is a functioning being, much the same as its parents. The individual, as Schrödinger put it, is 'reproduced without appreciable change for generations, permanent within centuries ... and borne at each transmission by the material structure of the nuclei of the two cells which unite to form the fertilised egg cell.' The only greater marvel, he said, was that the beings reproduced in this way were capable of gathering enough information about the process that they can understand it.

Schrödinger concluded that, in order to be stable and capable of sustaining life without throwing its future into chaos, a biological organism had to be much, much bigger than its constituent atoms and molecules. Quantum particles do strange things. They can be in two different states at once, or have their properties instantaneously altered by measurements performed on other particles. Such things could only be a hindrance to the smooth functioning of life's processes, Schrödinger said. Life, he decided, was built on a scale that would bury all the weird quantum effects. And that is why the size of a gene surprised him so much.

To find the size of a gene, biologists had to do things that should make modern-day lab workers thankful for all their technology. One estimate came through measuring the size of a fly chromosome, a component of every nucleus that was known to carry information determining a number of features. Something in the microscopic structure of the chromosome must carry the information, researchers argued, so the length of the chromosome, divided by the number of heritable features multiplied by the cross-section gave an estimate of the volume of the gene.

A second method for calculating gene size came from inspecting the chromosomes with a microscope. Picking the biggest available cells, those found in a fruit fly's salivary glands, biologists could see that the chromosomes within were striped. A count of those stripes gave a number that matched with what breeding experiments said were the estimated number of genes: 2,000. Each stripe, then, was roughly the size of a gene.

The estimates were crude, but they at least gave biologists – and interested physicists like Schrödinger – something to work with. The gene, they said, was roughly three hundred thousandths of a millimetre across. Schrödinger was fascinated by this figure. A gene of this size was only a hundred, or a couple of hundred, atoms in diameter and would contain 'certainly not more than about a million or a few million atoms'. That, he said, is far too small a number to turn the uncertainties of quantum behaviour into something 'orderly and lawful'. And here, he said, lies the marvel of life.

He was right for the wrong reasons. There is a marvel here, but the marvel is that evolution led the natural world to exploit quantum theory, not to bury it. As Carl Sagan was so fond of pointing out, terrestrial life is derived from the products of combustion in stars. The atoms forged in supernova explosions have, on Earth, gathered together to form molecular machinery that carries out the intricate operations we know as the processes of life. Life takes in fuel, stores it and burns it. Life senses the environment around it, and reacts and adjusts to the things its senses pick up. Life finds ways to reproduce itself. And somewhere along the way to creating these subtle and remarkable processes, evolution tried exploiting the quantum tricks available to molecules – being here and there at the same time, or existing in entangled states where distant molecules affect each other's states. A few of those tricks worked, and were retained.

Evolution has been cleverer than Schrödinger ever imagined. That is, for instance, the best explanation we have for why you react with disgust at the smell of a fart.

During the Second World War the Allies developed some odd weapons, but perhaps none as odd as its anti-personnel stink bomb. It was called 'Who Me?', and it was meant to make a soldier's clothes smell so disgusting that no one would be able to fight alongside him. The chemists couldn't quite get the formulation right, so it was never used in anger. But it is being weaponised again.

The US government has mixed the smell of Who Me? – a blend of rotting carcass and other pungent, sulphurous odours – with another of its proprietary odours: Standard Bathroom

Malodor. Government chemists came up with this hyperstrong excrement smell to test and certify the power of cleaning products. Who Me? and Standard Bathroom Malodor, when combined, are so unpleasant, so intolerable, that they will disperse a rioting crowd. The US military calls its concoction – classed as a non-lethal weapon – a 'disabling compound'. It plans to use the weapon, fired like an artillery shell, to make the most hardened, committed terrorist head outside and give themselves up.

No human being can tolerate the smell of rotting meat and excrement – and for good reason. Evolution equipped us with a sense of smell to help keep us alive. In nature, those smells are the product of bacteria that, consumed in sufficient quantity, will kill you. Even proximity to a bad enough smell will make you vomit, just in case you have ingested anything dangerous.

If there are smells we hate, there are some we love. 'Someday,' says Luca Turin, 'a scientist, or perhaps a poet, will manage to explain why we humans are so fond of ten-carbon (terpene) alcohols and aldehydes.' That's the scent of lavender, rose and lemon to you and me. The quote comes from his huge, hilarious and utterly absorbing book, Perfumes, written with scent expert and collector Tania Sanchez. Turin is a biophysicist, and is fascinated by the fact that we are so drawn to scents that evolution designed to lure pollinating insects – creatures to which we are, as Turin points out, 'not visibly related'. However, he was perhaps somewhat disingenuous in framing this conundrum, because he was already closing in on a solution. In fact it seems safe to say it will be a scientist, not a poet, that solves the problem, and it will almost certainly be Turin, who specialises in the science of smell (in academic corridors, it tends to be referred to as the far more respectable-sounding subject of 'olfaction'). The link between pollinating insects and we humans? Quantum noses.

Your nose has about 450 olfactory receptors – appendages

that hang down from your brain to help you turn a smell mol-ecule into a signal that tells you what is out there in your environ-ment. They all have different shapes, which is what led olfaction researchers to come up with the idea of a 'lock-and-key' mecha-nism for smell. The idea is simple: the smelly molecule is the key, and its shape means that it only fits into a particular receptor lock. Putting the key in the lock completes the circuit and sends a signal to the brain, which we interpret as a particular smell, the smell associated with that particular receptor.

It seems like a neat idea, but it has a fatal flaw. Humans can recognise around a trillion different smells. How is that possible with only 450 olfactory receptors?

Clearly, it can't be. It's not a burning question in science, however. The truth is, there's not a lot of research into olfaction compared to, say, vision. People who have lost their sense of smell – a condition called anosmia – are generally left to their own devices; it's not debilitating or common enough to merit a lot of attention, and there's not much the medical world can do anyway. What's more, the perfume industry is rather like the wine indus-try – it knows how to do what it does so well that the whole thing ticks along quite nicely without need of a smell scientist's input. An improved scientific understanding of smell is of very little use, really. And that means the subject has largely been left alone.

Turin wasn't the first to express frustration with this state of affairs. In the early 1970s several researchers pointed out that dozens of molecules smell the same but have very differ-ent molecular structures. Take the smell of bitter almonds, for example: there are at least seventy-five very different-looking chemicals that have this smell. You can't tell them apart with a sniff, but their molecular structures are extremely varied. Osyrol and β-santalol provide one example. They have completely dif-ferent shapes but both smell like sandalwood. If these chemicals don't all fit in the same receptor, they really shouldn't smell the same. Unless, that is, the lock-and-key theory is wrong.

Conversely there are molecules that have the same shape, and thus would fit in the same receptor, but smell very different from one another – vanillin and isovanillin, for instance. Physically, they are isomers, which means they have the same structure but are mirror images of one another. However, inhale a whiff of each, and you'd be amazed how different they smell. Vanillin was the downfall of the elm tree in Europe: the scent of vanillin in its bark attracted the beetles carrying the deadly fungus that has devastated them. Isovanillin smells much richer. Or try R-carvone and S-carvone. R-carvone has a spearmint smell. S-carvone carries the aroma of caraway seeds. Your nose isn't confused, even though they are mirror images of one another. So what is going on?

There is an explanation – and it precedes the lock and key idea. It came from a researcher called Malcolm Dyson, who was somewhat obsessed with smell, having been caught in a gas attack during the First World War. After that he played around with mustard gas and phosgene, using his knowledge of chemistry to change their atomic configurations. He noted that this could dramatically change the smell. Dyson put this down to molecular vibrations. He knew that the bonds between atoms in a molecule are springy, with each bond resonating at a particular frequency. And so in 1937 he declared that his nose must be detecting variations in the molecules' 'osmic frequencies'.

The idea didn't really catch anyone's imagination, but it seems reasonable. After all, other senses do the same. We distinguish colours because receptors in our eyes are tuned to different wavelengths of light: the waves hit our retinas and we distinguish between them because of the different frequencies at which their electromagnetic fields oscillate. Hearing, too, relies on oscillations – this time in the pressure applied to the hairs in our inner ear by incoming sound waves. A soprano hits us with fast oscillations; a baritone's voice is composed of slower vibrations in the air. In both cases, humans are equipped to hear only a limited set of the possible oscillations. Other animals hear and see at different frequencies. So why shouldn't the same be true of smell?

Dyson's theory becomes even more plausible when we acknowledge, for instance, that only two things in the world smell of rotten eggs: thiol, composed of an oxygen and a sulphur atom; and boranes, which are composed of boron and hydrogen. They have nothing in common except an exactly identical molecular vibration frequency. As Turin says, it's surely 'too marvellous to be a coincidence'.

Sadly, however, no one took much notice of Dyson. When someone proposed the lock-and-key mechanism for smell, which was a little more tangible than vibrations, it quickly became the frontrunner. Turin, however, is convinced that the lock-and-key idea isn't opening anything up. It's time, he says, to go back to Dyson's vibrations – but with a quantum twist.

Turin's stroke of genius was to look for a pair of molecules with structures that were identical in almost every test you could apply, but would somehow smell different. He found what he was looking for in acetophenone and acetophenone-d8.

These two compounds are even more similar to each other than normal isomers. Even a chemist couldn't tell the difference between the two. If you were able to look at them under a super-powered microscope, you wouldn't see a difference. Their molecules differ in only one, extremely subtle respect. The hydrogen atoms in acetophenone have just one proton in their atomic nucleus. The hydrogen atoms in the acetophenone-d8 are heavy hydrogen, or deuterium: that is, they carry a proton and a neutron, and hence the molecule is said to be deuterated.

It is about as subtle a change as can be achieved, but it has consequences – especially when it comes to vibrations. We can look at the constituents of a molecule and work out the natural frequency at which its bonds will vibrate. Normal acetophenone has a carbon–hydrogen bond, for instance, which wags at a particular frequency. Deuterated acetophenone has the same bond, but, because of the neutron in the hydrogen atom, it vibrates at a different frequency.

That neutron, in fact, makes the bond vibrate at the same frequency as a bond between a carbon and a nitrogen atom. Which makes it terribly interesting that the deuterated acetophenone smells more strongly of bitter almonds; benzonitrile, which is dominated by carbon–nitrogen bonds, is a prime candidate for anyone looking to create a bitter almond smell.

There are other differences in smell between deuterated and undeuterated acetophenone. The neutron makes the deuterated version fruitier to Turin's trained nose, with a 'less toluene-like character than acetophenone'. However subtle and subjective the effects of deuteration might seem to us, though, they hit fruit flies like a sledgehammer. In 2011 Turin and his colleagues published a paper describing how they had selected the flies as 'unbiased and objective subjects' to distinguish between the vibration and the lock-and-key mechanisms. The flies could clearly tell the difference not only between isomers, but also between normal and deuterated versions of acetophenone. The conclusion was clear: 'Flies can smell molecular vibrations.' The really interesting thing, though, is that the vibrational smell idea only works because of quantum theory.

Quantum rules say that the electrons within an atom can only hold certain amounts of energy. If you want to fire an electron to take a signal into the brain, for instance, you have to give it a precise dose of energy – no more, no less. The same rule holds for the photoelectric effect, for which Albert Einstein won his Nobel prize. He showed that electrons can be freed from the surface of a metal only by light of certain distinct energies. Get the photon energy wrong – higher or lower – and there'll be no current, no matter how many photons hit the surface. We also know this quantisation, or signal processing, effect occurs in the vibrations of bonds within a molecule. They need a certain minimum amount of energy to move. Bang them with less than the minimum energy, and they will stay resolutely still.

Turin's suggestion is that these quantum restrictions mean that a molecule lodged in an olfaction receptor will only switch the receptor on if there is a resonance between the vibrational energy of its bonds and the energy required to fire a signal from the receptor to the brain. That way an electron can perform a quantum 'tunnelling' manoeuvre between the molecule and the receptor, starting a cascade of signals the brain interprets as a particular smell.

It's still, to smell researchers at least, an unproven idea. The barrier is set high in science, and before the suggestion of a quantum mechanism for smell is accepted (being the only idea that fits all observed facts isn't good enough), Turin will have to take an olfactory receptor and demonstrate its quantum proper-ties – that it can be stimulated by vibrations of one frequency, but not another. It's a huge ask, but Turin is working on it.

It's worth restating, though, that without this quantum expla-nation, we have no understanding of smell at all. That's exhibit one in the case for a quantum-based biology. Look out of a window, and you'll almost certainly see an example of exhibit two: the humble leaf.

If you get the chance, read Sara Maitland's beautiful and startling story *Moss Witch*. It depicts an encounter between a botanist, whose beard is 'the colour of winterkilled bracken' and a strange, magical creature who has the mosses of the wood at her command.

The Moss Witch utters an achingly beautiful line during her conversation with the botanist. She is native to the wood where they meet, and tells the botanist of her sadness that evolution has left her alone and unable to create company for herself, even by cloning. 'That's one of the problems of evolution – losses and gains, losses and gains,' she says. In the end, Maitland tells us, mosses – and moss witches – learned to accept their lowly place in the natural world. 'Moss witches, like mosses, do not compete; they retreat.'

Evolution's losses and gains led mosses to retreat in the face of something we take for granted now: the leaf. It is the source of life to us, because it offers a form of stored energy that we can process; without the leaf, it's hard to see how we and the other animals would have arisen in the evolutionary heritage of the earth.

The leaf, it turns out, is a remarkable piece of technology. Light lands on a leaf in quantum energy packets called photons. Every second it spends in bright sunlight brings a leaf thousands of trillions of photons. And almost every one of the red ones is captured and its energy processed.

Let's trace the path of that energy through the plant. The first stop is the antenna, an assembly of hundreds of thousands of chlorophyll molecules. Hit by the photon, an electron within the chlorophyll will become 'excited' and use the extra energy to bounce through the antenna to a bridging area. This is composed of a complicated arrangement of chlorophyll molecules: you can imagine it as an intertwined maze of rope bridges that, if you can find the path through, lead to the all-important 'reaction centre'.

Once at the reaction centre, the electron creates a permanently separated charge that is, essentially, energy that can be stored. It's the same thing that makes a fully charged battery such a useful source of power. At that point, the energy is safe and properly stored, ready for use by the plant. There's just one catch: the reaction centre is a nearly impossible place to reach.

Getting through the bridging area is a nightmare, and it has to be done quickly. The maze of wobbly rope bridges will absorb all the photon's energy if it takes too long to find the reaction centre; a packet of energy has about 1 nanosecond to get across the bridge. Some might chance upon the right path and make it through in time; most, though, won't. This is where the quantum tricks – and Gregory Scholes's research team – come in.

Scholes is no scruffy botanist; there's no beard the colour of winter-killed bracken on his face. Look at his research group website, and you'll discover that Scholes wears Prada shoes and a 'porn-star shirt'. Feedback from undergraduate students features cheers ('Eyecandy'; 'Hurray for Dolce & Gabbana and phys chem!') and jeers. One student says Scholes 'would be better off teaching a fashion class than chem … I couldn't get past the fact that his pants were way too tight.' As a lecturer, it's not a ringing endorsement. But there's no doubting that his research is making waves.

The group's 2010 *Nature* paper doesn't say whether Scholes himself went down to the Toronto beachfront and pulled some green slime from Lake Ontario. Maybe he sent someone else to do the dirty work. Either way, when they got the algae back to the lab, they extracted the parts of the plant that convert light to stored energy. Then they shone laser pulses on it to mimic a hit from sunlight and timed the passage of the energy through the system. Their conclusion? 'Our result suggests that the energy of absorbed light resides in two places at once.'

This weird phenomenon is officially known as quantum super-position. It is written into the mathematics of quantum theory, which describes the position of a particle only in terms of the probabilities of finding it in various different places. In certain circumstances, the particle can occupy multiple positions at once. And it turns out that, to achieve the remarkable efficiency of photosynthesis, algae and their evolutionary descendants, plants, employ quantum superposition so that the energy can simultaneously travel along all possible paths to the reaction centre. That way, it always gets there on time.

As far as we understand it, when a photon of sunlight hits a leaf, its energy is passed to electrons within the leaf's chlorophyll molecules. Those 'excited' electrons move to the reaction centre not by random hops, as was thought, but by exploiting their quantum mechanical ability to take all available paths at the same time. A similar trick has been found in some bacteria, such as *Chlorobium tepidium*. They too, seem to use quantum superposition to deliver energy to the reaction centres of their photosynthetic apparatus.

It's such an exciting discovery because it raises the possibility that we could do the same. Part of the reason for leaf-wielding plants' domination over mosses is that they have so much more equipment carrying out photosynthesis. If the messy world of biology can exploit this quantum mechanical trick to the maximum by evolutionary trial and error, perhaps we can deliberately engineer the same feats to occur in our solar panels.

Despite researchers' best efforts, we have been stuck with low-efficiency conversion of light to electricity for far longer than we expected. The current generation of photovoltaic panels are able to convert roughly 20 per cent of the light falling on them into electricity. If you're willing to pay \$100,000 per square metre for your solar panels, you can get up to around 35 per cent efficiency. It is hard to measure the efficiency of a plant's conversion of light to electricity – that is, moving electrons – accurately, but it seems to be in the region of 90 per cent. And that's in weak sunlight. What's more, evolution hasn't had to pay a dime.

It's worth pointing out that photosynthesis is not converting 90 per cent of the light that hits a leaf to stored chemical energy. Looking at it that way, the process is only around 5 per cent efficient. The point is, though, the potential is there to do much better: in general, plants won't store more energy than they need, and in strong sunlight the potential is so high that they have even evolved mechanisms for dumping some of the energy, so as not to overheat.

It was insightful of Erwin Schrödinger to reason that the biological cell must work on large enough scales to avoid the downsides of quantum theory. However, it never occurred to him that the upsides of quantum weirdness might actually be useful – and essential to us all. It seems to be what makes plants able to take the energy of the sun and store it within their cells. That stored energy is the basis of life on Earth: almost every living creature is dependent upon the plants' harvest – a harvest that itself depends on quantum weirdness.

I know what you're thinking: it's still not as weird as a robin wearing an eye patch. It's time to go back to Wiltschko's birds.

If you want to imagine what Wolfgang Wiltschko looks like, just think of Santa Claus – which is appropriate, considering his long association with robins and winter. Wiltschko published his first major article showing that robins migrate south in winter using a magnetic sense in 1965. However, the one that Wiltschko himself references is in 1968. He wrote and published it in his native German; the title roughly translates as *On the Influence of Static Magnetic Fields on the Migration Orientation of the European Robin.*

By messing about with his artificial fields while a robin sat in the Emlen funnel, Wiltschko was able to make a slew of surprising discoveries. First, the robins' compass is not like the compasses we know. Ours point to the magnetic north, but the robin is equipped with a compass that senses how the magnetic field is orientated relative to the Earth's surface. To go north it follows the field in the direction towards which a constant magnetic intensity leads it downwards. To move towards the equator it moves in the direction in which a line of constant intensity – rather like a contour line on a map – rises away from the Earth's surface.

Next he noted that the robin can only sense this magnetic field within a small window of intensity. Make the field too weak, and its ability to orientate itself disappears. Even more interestingly, that's not to do with some threshold below which its senses can't pick up the field. Make the field too strong, and the same thing happens.

Then the Wiltschkos noticed that the robin needs a little bit of light to orientate itself. And it is light of a particular colour: a blue-green hue. The final, and truly bizarre observation – and this is where the eye-patch comes in – is that this light has to come in through the bird's right eye. Cover the left eye, and the bird can still navigate. Swap the patch to the right eye, and it is lost.

This trail of clues led a trickle of quantum physicists to the Wiltschkos' door. So far, they are the only ones with a coherent explanation of all the Wiltschko's observations. And that is how, in 2010, the Wiltschkos ended up travelling to Brussels with Thorsten Ritz.

Ritz, too, is German, though he now works at the University of California, Irvine. He, Wolfgang and Roswitha were in Brussels to attend the 2010 Solvay Conference in Chemistry. It is a hugely prestigious meeting, with a history that goes back to 1911, when scientists such as Albert Einstein, Henri Poincaré and Marie Curie gathered to discuss the need to reconcile classical and quantum physics. It was at the fifth Solvay conference that Einstein expressed his distaste for the statistical nature of quantum theory by stating god does not play dice with the universe.

The 2010 meeting was to discuss not a problem, but an intriguing possibility: 'Quantum Effects in Chemistry and Biology'. At the chair's invitation, Ritz and the Wiltschkos laid out their observations, ideas and arguments in the Excelsior Room of the ornate Hotel Metropole. The best explanation for bird navigation, they said, involved quantum mechanics.

To explain, we need to start with the phenomenon of phosphorescence. Perhaps you, or someone in your house, has a set of glow-in-the-dark stars that emit a faint, eerie glow long after the lights have been turned out. This is phosphorescence, and it works through quantum rules.

Storing up the energy of light, and releasing it slowly, is something an atom can only do if it has a certain configuration of electrons surrounding its nucleus. The process requires one of its electrons to use the light energy to go into what's known as a 'triplet' state. It's as if the electron uses the energy to climb up onto a narrow beam and start walking along it. You know it will eventually fall back to the ground – emitting a little light energy as it does – but it won't happen straight away. The length of time before that electron falls is random, the same randomness that decides the emission of radioactive particles from a lump of uranium. The probability of such a 'decay' within a particular time is determined by the equations of quantum theory.

What's going on in the robin's eye is, according to Ritz's theory, just like phosphorescence. The energy in that blue-green light – and *only* that blue-green light – is just right to kick an electron into a triplet state. Where does that electron come from? From cryptochromes, protein molecules embedded in the bird's retina that react to electromagnetic radiation by firing electrical signals. Plants have them too; they enable flowers to follow the sun, and certain species also use cryptochromes to respond to magnetic fields.

The electron doing the work in the cryptochrome originally exists as part of a pair, known as a 'radical pair' because they make their molecule extremely reactive. When one electron gets kicked up by the blue-green light into its triplet state, the two electrons become physically separated, existing in slightly different positions on the molecule. However, they remain correlated, or 'entangled'; do something to one, and it changes the state of the other. The result of this, in a magnetic field, is that they exist embedded in very slightly different magnetic field intensities and orientations. That creates a kind of quantum 'strain' between them. It is this strain that might give the bird its magnetic sense.

Early takes on this idea thought that the electron in the excited state quickly 'decayed' back down to its original place, giving out a pulse of energy that the bird converted into some signal. However, experiments have revealed that the radical pair lasts for an extraordinarily long time: some tens of microseconds. That means the pulses of energy wouldn't come often enough to be useful; instead, the best explanation seems to be that the long-lived strain between the two electrons gives the robin an ongoing and enhanced experience of the world. The current conjecture suggests that the birds might see the magnetic field as a distortion of their normal vision.

You might be able to get a feeling for this if you can find a blank, white sheet of paper and a filter that polarises light. Alternatively, pull up a blank word-processing document on your laptop screen – LCD screens have a polariser built in. The way the lutein molecules that sense blue light are arranged at the back of your eye means that you can sometimes detect polarised light in the same way a robin might detect a magnetic field. Staring at white polarised light can – for some people, at least – create a yellow bow-tie shape in their field of vision. The illusion is called Haidinger's brush. If you turn the screen or the polariser through 90 degrees, you'll see the bow tie turn with it. It's superimposed on your vision, like the kinds of heads-up display such as pilots and some expensive satnavs use. Somewhere in evolutionary history, our forebears lost the ability to sense magnetic fields, and we have almost no ability to detect polarised light either. Other animals, though, have retained these supersenses as a means of – in this case, at least – surviving the harsh conditions of winter. Despite Schrödinger's best intuition, evolution seems to have put quantum theory to work.

It's an exciting possibility. We mentioned earlier that this dis-covery might help us build better solar panels. It might also help us towards the goal of building usable quantum computers. These ultra-powerful machines, which we'll explore in more depth in a later chapter, rely on exploiting the quantum phenomena of superposition and entanglement. That, though, is a lot easier said than done.

We have always struggled to control the quantum world; it only comes to heel in carefully designed labs, where the quantum systems are isolated from every possible disturbance. Vibrations caused by passing traffic have to be filtered out, and the quantum calculations only work when the atoms involved are held at ridiculously cold temperatures, close to absolute zero (–273 °C), where molecular vibrations are stilled and nothing has the energy to disturb the delicate quantum states of matter entangled together and performing the feats of superposition. We devise cunning ways of looking at the quantum circuits so as not to cause the system to lose information that would collapse the superposition before the computation is complete.

And yet the robin wends its merry way from Europe to North Africa and back without any of this. If the bird does use quantum tricks to navigate its way around the world without the need for bulky refrigerators or clever suspension systems, the next decade of research into quantum biology might just get us further than clever engineering ever has. For now, though, let's move on to a different impact of quantum theory on biology – the one that reveals that we are nothing more than holograms projected from the edge of the universe. Hold on to your mind: our new view of cosmic reality is going to give it a good shake.

8

THE REALITY MACHINE

Our universe is a computer, and we
are the programmers

We are a way for the cosmos to know itself.
Carl Sagan, Cosmos

Occasionally, on a sultry night in Singapore, Vlatko Vedral can be found outside a whisky bar in Clark Quay, a single malt in the glass on the table in front of him, a cigar glowing in his hand. The quay is vibrant after dark, crammed with young people and professionals newly released from the tall glass and steel palaces where they work. It is beautifully lit and – like the rest of Singapore – beautifully clean and wholesome. Children allowed to stay up late will play, laughing and screaming as the nozzles set into the circular centrepiece of the district shoot water jets out at unpredictable moments. Music leaks out from bars, punctuated by the occasional shrill shriek from the people braving the enormous bunjee cradle that flings them at heart-stopping speed towards the heavy sky.

Vedral, a tall, thoughtful Serbian-born physicist, watches the scenes from the semi-darkness outside the Highlander Bar. You can be sure that, if he is awake, Vlatko Vedral is usually thinking. The thing he thinks most about, as the dessicated cells in the tobacco leaf slowly burn away, is information.

'The laws of Nature are information about information and outside of it there is just darkness. This is the gateway to understanding reality.' That is Vlatko Vedral's take on the universe. Information, he says, is a far more fundamental quantity in the universe than matter or energy. This has extraordinary implications for every one of us, Vedral says, because we are processing, synthesising and observing that information in order to construct the reality around us.

In Vedral's mind, information is everything and everything is information. Even that cigar: if he wanted to describe the leaves it contains he would have to transmit a lot of information about their intricate structure. Burning the leaves is the equivalent of erasing the information about how its cells were put together. It increases the disorder of the universe. In Singapore, one of the most carefully ordered cities on the planet, Vedral's cigar-smoking is a little act of defiance.

Vedral's interest in information started with a 1994 encounter with three words: 'Information is Physical'. When we think of information, we tend to think of something abstract: a test score, perhaps, or a piece of news. But that news is somehow encoded by the neurons in your brain. To get there it was encoded as writing – ink on paper, perhaps, or pixels of liquid crystal on an electronic screen. Trace the history of that information, and it was never abstract, always encoded. It was once purely electronic, carried in a pattern of electrons moving through a circuit on a microchip inside a computer. Before that, it was transmitted along optical fibres, written into the properties of a beam of laser light. Information is physical. Always.

'Information is Physical' was the battlecry of a man, now deceased, called Rolf Landauer. Landauer is also known to physicists for his observation that information gets wiped. The light that carried your news no longer exists; its energy has been converted to another form, and all the information is gone. That process of erasure, Landauer said, costs energy, and contributes to the universe's descent into disorder. This is Landauer's principle. It applies to the tobacco leaves smouldering away in Vlatko Vedral's hand, and to the thoughts roaming through his mind. It's about what happens, as in Landauer's case, when a human brain is ravaged by cancer. It's about the outcome when a million stars fall into the inescapable clutches of a black hole, and when the ice melts in a whisky glass. Everywhere, all the time, information is being lost. This is how some of our finest minds came to a beguiling conclusion about who we are, and where we are. We are, they say, inside a vast computer that we call the universe.

In the south-eastern corner of the constellation Sagittarius, hidden behind clouds of interstellar dust, is a monster: the black hole known as Sagittarius A-Star. We can't see it, as such. We only know it is there because of what it does to the stars in its vicinity. One has been seen moving at 5,000 kilometres per second in a circular orbit around this blank area of the Milky Way. There are around a dozen such objects, all orbiting the same region of blackness in a way that suggests whatever is in the small region of darkness 26,000 light years away has an enormous mass.

Black holes are astonishing things. They are so dense that their gravitational field pulls in everything and anything that comes too close. There is no escape for matter or light that has crossed the 'event horizon', the point of no return. Sagittarius A-Star is a particularly big one: a supermassive black hole. It has a mass around 4 million times that of our sun. Over its lifetime it will swallow millions, perhaps billions of stars and perhaps planets too, all drawn by its inescapable gravitational pull. It's strange to think, then, that this monster won't always be there.

In 1974, the same year Sagittarius A-Star was discovered, Stephen Hawking conducted an imaginary experiment. He combined everything we knew about black holes with everything we knew about quantum theory, and came up with a discovery that surprised everyone.

One of the defining traits of quantum theory is that things don't have a definite existence, but are 'probably' there, should you happen to look. In a quantum world, you might put this book down on a table, only to find that the table isn't there any more, and the book falls to the floor. It's a pretty improbable event, but it's still possible. Just as possible as the disappearing table is the idea of particles that spontaneously pop into existence from nowhere. It sounds ludicrous, but the 'uncertainty principle' that rules quantum theory says it can and does happen. All over the universe, and right in front of your nose, right now, pairs of particles – a particle and its antiparticle – are popping into existence for a tiny fraction of a second, then annihilating each other to leave their little region of empty space empty once again.

Hawking imagined what would happen when those pairs of particles popped into existence on the event horizon of a black hole. Most would meet and annihilate. But there might be a pair that straddles the event horizon. The one inside the event horizon would be pulled in to face its doom at the centre of the black hole. The other, beyond the horizon, would float away, free. The black hole would have, effectively, emitted a little particle.

Hawking knew this was a revolutionary notion. If the black hole emits a stream of particles over its billions of years of life, the energy for those particles has to come from somewhere. The black hole, he reasoned, would be depleted of energy. That would make it shrink. Eventually, he said, a black hole will evaporate away to nothing.

If Sagittarius A-Star will slowly disappear, that raises another beguiling question. When it is gone, what will have become of all the information it swallowed? It's a question we're still trying to answer.

'Stephen, as we all know, is by far the most stubborn and infuriating person in the universe.' That was how Leonard Susskind decided to introduce his talk celebrating Stephen Hawking's sixtieth birthday.

The two men first met in the house of Werner Erhard, a strange meeting of minds if ever there was one. Erhard was 'America's first self-help guru', according to the *Financial Times*, famous for bringing celebrities such as Cher, John Denver and Diana Ross into his programme for dealing with life's difficulties. Erhard's thinking was inspired by Zen Buddhism and focuses on working with what you have, rather than worrying about what you haven't. He coined the phrase 'thank you for sharing' to encourage and reward those who offer their thoughts and insights for the greater good.

Erhard is a fan of physics. He liked to gather physicists so he could hear them talk to one another, perhaps hoping to absorb some scientific wisdom. And, as it turned out, he inspired some. In 1981 Hawking and Susskind were among the guests at one of Erhard's gatherings. The two men got talking about Hawking's public declaration that information would just evaporate with the black hole. Hawking hadn't just declared it: he had proved it mathematically. It was a proof that Susskind refused to accept. He told Hawking so, and Hawking smiled and declared himself right anyway. It's unlikely that Susskind, a fierce-eyed, bearded furnace of intelligence, thanked Hawking for sharing.

Susskind spent the next two decades on a crusade to prove Hawking wrong. The result of that crusade has been the discovery that our universe is ephemeral, a ghostly illusion. Yes, the table can disappear from existence in improbable situations. But that's not all. It turns out that the table – and everything around you – is a hologram.

Holograms are rather tricky things to grasp – in every sense. They trick the eyes into thinking an object is real by harnessing certain properties of light, and they befuddle the mind when we try to understand how they are made without reference to the mathematical properties of light. But, put simply, holograms gather all the available information about how far away different points on an object are from each other by recording how much further light has to travel to your eye from one point than it does from another. All that information is encoded into the light emitted from the lasers that create the hologram. The interaction of those laser beams creates the illusion that the object is in front of you.

When you think of a hologram, you probably think of the moment in *Star Wars* when the robot R2-D2 projects a hologram of Princess Leia. The thing is, we all know the hologram was only a representation of the princess. If we, and all our reality is a hologram, where is the projection coming from? Where is our R2-D2? And where is the real us? The answer to that question – at least the one that changed Stephen Hawking's mind – came from an unassuming Argentinian physicist called Juan Maldacena.

For years Hawking and Susskind had argued from opposite

viewpoints with no compromise on either side. Hawking would infuriate Susskind with a taunt worthy of the playground: 'I'm right and you're wrong,' Susskind says. In a 2008 interview with the *LA Times*, he said, 'I love the man, but I wanted to grab him by the neck and shake him a little bit.' And then Juan Maldacena broke the deadlock.

Maldacena works on string theory, an attempt to use mathematics and geometry to show how pure energy can provide the building blocks of a universe. The hope is that string theorists can build a theoretical universe that looks and behaves exactly like ours. If they can, it might tell us something about our origins.

In 1997 Maldacena made a big breakthrough towards this – though perhaps it doesn't sound like it at first. He created a working string theory universe with five dimensions of space; two more than the three we inhabit. He then showed that in this five-dimensional universe the laws of physics worked the same way as they would in a different universe, one that had only four dimensions of space. The intriguing thing was that, just as the two-dimensional surface of a sphere encloses a three-dimensional object, those four dimensions provided the shell that enclosed the five-dimensional universe.

If the laws of physics on the surface were equivalent to the laws of physics within the object, all the information about the object is held on its surface. It's like discovering that the skin of an apple contains all the features – the pips, the flesh, the core – of the middle. In other words, you can hold all the features and characteristics of an object on its outer surface.

It must have been somewhat annoying for Susskind – also a string theorist – to have Maldacena change Hawking's mind. After all, Susskind had been arguing along the same lines for years. He was one of the first people to suggest a link between the physics inside a black hole and the physics on the surface, that the amount of information in the interior is proportional to the area of the spherical event horizon that delineates its point of no return. He had even talked about how the information escaping through Hawking radiation might leave an imprint on the event horizon itself, getting rid of that troublesome loss when the black hole evaporates away. But Maldacena's mathematics did what Susskind's arguments couldn't. In July 2004 Hawking sent a note to the organiser of the 17th International Conference on General Relativity and Gravitation, due to be held in Dublin later that month. It said, 'I have solved the black hole information paradox and I want to talk about it.'

That was, perhaps, something of an over-statement. He had built on the work of many others, and the solution he presented in Dublin was not complete. But we now have a solid grasp of how black holes preserve information in their event horizons. What it teaches us is that the same principle may be creating what we call reality. All of the information that makes up our existence – the structure of a tobacco leaf, the way stars are distributed in the Milky Way, the genetic recipe for Leonard Susskind's beard and the program embedded in Stephen Hawking's voice synthesiser – exists on the boundary of our universe. And, in the shape of an unexpected result from a physics experiment in the German countryside, we might even have some supporting evidence.

On 4 September 1995, computer programmer Pierre Omidyar founded a small company called AuctionWeb. On the same day, a large red excavator rolled into a field in Germany and began digging a trench that might have more impact on humanity than AuctionWeb – now known as eBay – could ever hope to achieve.

The trench was dug for GEO600, a joint British–German effort to look for ripples in space and time. Their existence is predicted in Einstein's general theory of relativity; it says that cataclysmic cosmic events, such as the merger of two black holes in a distant galaxy, should create a ripple in the fabric of the universe. No one has yet detected a gravitational wave on Earth. But GEO600 has seen something far more interesting than a cosmic vibration. It has, it appears, detected the true nature of reality.

GEO600 is not a huge facility, like CERN's Large Hadron Collider. Anyone who came across it would see little more than a huddle of steel-grey cabins in the countryside south of Hanover. The impressive stuff is in underground tunnels, two arms that stretch out for 600 metres at right angles to one another. When the detector is operating, the tunnels carry laser beams that allow its research team to detect tiny wobbles in space. That's because the instrument measures space with a sensitivity equivalent to a ruler that can determine the diameter of a proton. It is sensitive to the traffic rumbles from nearby roads, and it can even tell when a large cloud is passing overhead. Operating the machine successfully involves making sure none of these sources of 'noise' can disturb the measurements. Unfortunately, when they got it working, the detector was plagued by a noise no one could account for. Something was creating a signal where there should be none, and nothing could make it go away.

The noise was initially rather frustrating to Karsten Danzmann, who runs the detector – he thought it was an indication of some fault in the equipment. But then Craig Hogan got in touch.

Hogan, the director of Fermilab, the premier laboratory in the US for exploring fundamental physics, had been teasing out the implications of the holographic principle. He was wondering what it meant that all of the contents of our universe result from something encoded on its outer shell. His first conclusion was that there can be only so much information to go round. If you want to stud the universe's skin with bits of information, there's certainly a lot of space. But there's not an infinite amount. And that means the universe itself can't be infinitely detailed. Like a photograph, you can't zoom in on it forever without finding that things get grainy. Hogan did some calculations and found exactly where the pixels of the universe would start to show up. Then he looked around for instruments that were measuring things at that kind of level. That's when he came across GEO600.

The blurriness of space and time has different effects, depending upon the way you're trying to measure it. GEO600 makes measurements using laser light reflected off mirrors at each end of its 600-metre-long arms. Those reflections result in light that is spread over a range of frequencies. Hogan worked out how the limited information on the boundary of the universe, pixellating space and time, would create a blur in the exact position of the mirrors. The blur would, he determined, give a different intensity of light at each frequency in the range. That creates a 'power spectrum'. Hogan sent his power spectrum to Danzmann, who wrote back, astonished. It was his noise.

Or, at least it looked like his noise. It was close enough to get both Hogan and Danzmann excited, but we're still not sure – which is why Hogan is constructing a purpose-built holographic universe detector at Fermilab. He wants to know for certain if we're all just a projection from the edge of space and time. We don't yet have the answer.

It may just be worth pausing here for a moment. We have come from recognising a fundamental principle in the universe – that information is held in physical things, but continually being displaced and distorted – to find a startling new perspective, one in which every bit of information is being projected from some point on the outer skin of our universe. The basic, original information is unchanged; it is our view of the information that shifts and changes as the cigar burns down and the black hole swallows another star. We are watching a movie in which we are, from our perspective at least, the principal actor. The projector for this movie is shining in from the edge of the universe. So, what's this movie about? Computing, that's what.

It's not an unprecedented idea. In *The Hitchhiker's Guide to the Galaxy*, for instance, Douglas Adams portrays our planet as part of a huge computer designed to find the answer to the question of Life, the Universe and Everything. It seems Adams's vision may have been too small, though. If the biggest brains on our tiny planet are right, the whole universe is a computer.

Maybe Adams thought we weren't ready for the idea. He certainly knew about it; *The Hitchhiker's Guide to the Galaxy's* computer, Deep Thought, was inspired by an Isaac Asimov story, *The Last Question*, in which humans seek to reverse the inexorable increase of disorder in the universe. *The Last Question*, published in 1956, is a masterpiece: it begins with a $5 bet over highballs, and ends with humanity absorbed into the vast intelligence of the cosmos, all wrapped up in a book that takes around half an hour to read. It is by far Asimov's favourite of all his writings, he said.

1956 was also the year a young man called Edward Fredkin first encountered computers. Asimov can lay claim to the original notion that the universe is a computer, and a German computing pioneer called Conrad Zuse made the first forays into how such a universe might arise from mathematical considerations. But it is Fredkin, a multimillionaire owner of an island in the Caribbean, former fighter pilot, and ex-professor at the Massachusetts Institue of Technology (MIT), who is now widely considered the architect who produced the blueprint. And it's quite a drawing.

In Fredkin's conception there is no such thing as space, or time. There are no fundamental particles – no electrons or photons or neutrons or quarks. All that stuff is a result of tiny pixels of information, something like a three-dimensional version of your TV screen. Instead of TV signals switching the pixels on or off, however, it's the neighbouring pixels that determine what state each pixel will take.

Just as your TV's pixels can create an image of a face or a planet, Fredkin's fundamental pixels of information can create particles or forces. The difference is, it's not just an image of a particle or a force. It's the real thing.

The way those particles behave is a result of the rules that switch neighbouring pixels on and off. So the pixels that make up the negative charge of an electron will follow a rule that makes the 'electron' configuration move towards the pixels that make up a positively charged proton. Not that the pixels themselves move, any more than the lights in your TV move. It is simply the pattern of 'on' and 'off' pixels that moves, just as the undulations of a Mexican wave in a crowd come from a pattern of people standing up or sitting down. From this humble set-up, Fredkin says, comes all of physics.

He calls it 'digital physics', since it arises from the two distinct states of his pixels. Nothing in Fredkin's universe is continuous: it is all binary. It's no coincidence, he says, that progress has moved us into an era when we use the same fundamental idea to perform most of the day-to-day tasks of civilised humanity. If that's how the universe works, it's almost inevitable that we would eventually come to structure things the same way.

In his mind, at least. It's worth stating at this point that Fredkin is not revered as a genius by everyone working in science. Most people have never heard of him. Of those who have, most people consider him an eccentric with interesting but flawed ideas.

In 1974, for instance, the revered physicist Richard Feynman brought Fredkin to the California Institute of Technology. He wanted to absorb everything Fredkin knew about computing and see if he could apply it to physics. 'If anyone is going to come up with a new and fruitful way of looking at physics, Fredkin will,' Feynman once said. But in the end, even Feynman couldn't buy into Fredkin's vision.

Fredkin has always been an outsider. At school he was a misfit, always the last one to be picked for any team. Each side would fight *not* to have him on their team; they would rather be a man down. In the kingdom of the unpopulars, he was the anti-king. 'I was in the pole position. I was *really* left out,' Fredkin recalled in a 1988 interview for *The Atlantic* magazine.

It's still true today. That may be because, like Einstein – and unlike Feynman, Susskind, Vedral and countless others – Ed Fredkin could never stomach quantum theory.

Einstein's big problem with quantum theory was that it relies on probabilities. It has uncertainty built right in. This is the root of his famous complaint: it simply can't be, he said, that God 'plays dice with the universe'. Fredkin feels the same way: if we observe quantum systems following the laws of probability, that can only be because we haven't yet discovered all of the principles that dictate their behaviour.

The trouble is that scientists can only go by what the universe tells them, and the universe has so far been entirely unequivocal about the role of randomness and uncertainty. There is almost no one who works with fundamental materials who doesn't accept that things are fundamentally fuzzy. And that must include the way the universe processes information.

Fredkin's cosmic computer is based on certainty: its pixels are either on or off. But we know the universe doesn't work quite like that. Everything, from the electron to the atom to the forces that make them do what they do, follows the probabilistic laws of quantum theory. So, if the universe is a computer, it can't be a normal computer. The pixels are not on or off: they are a bit of both. And that makes the universe a quantum computer.

The first person to suggest the universe is a quantum computer was a pony-tailed MIT physicist called Seth Lloyd. Lloyd is a fun person to be around; jolly, always offering up little jokes and asides, laughing in a goofy, ululating tone. He does have a serious side, though, as he ably demonstrated at the end of his astonishing book *Programming the Universe*.

In the book's epilogue Lloyd tells the story of how he watched his friend and mentor Heinz Pagels die. In 1988 the two men were climbing Pyramid Peak, a 14,000-foot mountain in Colorado. Thanks to an ankle weakened by childhood polio, Pagels lost his footing and fell. As Lloyd looked on, helpless, Pagels plummeted into a deep gorge.

It would have been a particularly haunting moment for Lloyd to watch because Pagels had described a premonition of his death in his own book *The Cosmic Code*: 'Lately I dreamed I was clutching at the face of a rock but it would not hold. Gravel gave way. I grasped for a shrub, but it pulled loose, and in cold terror I fell into the abyss…'

Having imagined his death, however, Pagels also imagined his immortality: 'What I embody, the principle of life, cannot be destroyed … It is written into the cosmic code, the order of the universe. As I continued to fall in the dark void, embraced by the vault of the heavens, I sang to the beauty of the stars and made my peace with the darkness.'

Lloyd reflects this in his elegy to Pagels. 'We have not entirely lost him,' he writes. 'While he lived, Heinz programmed his own piece of the universe. The resulting computation unfolds in us and around us.' It is a beautiful, haunting sentiment, but to Lloyd it is more than that. It is the culmination of his life's work so far.

Seth Lloyd's star has always shone bright. Rolf Landauer once offered him a job; Lloyd turned Landauer down because he also got an offer from Nobel laureate Murray Gell-Mann. Subsequently Lloyd came up with the first design for a quantum computer that really could be built. The idea behind the machine is rather simple. Instead of using standard electronic circuitry to store and process information, the quantum computer encodes information in a particle that follows the laws of quantum theory. Where a standard circuit would be either off or on – creating a 0 or 1 in binary digits – the quantum nature of the particle means it can be 0 and 1 at the same time. This is a phenomenon called superposition, the same peculiar ability that allows plants to channel light energy through many paths at once.

Another purely quantum phenomenon is 'entanglement'. This allows particles that have been suitably prepared to share a seemingly telepathic link. A measurement carried out on one particle will alter the result of a subsequent measurement of its entangled twin, even if they are on opposite sides of the universe. When Einstein saw this in the equations of quantum theory, he dismissed it as 'spooky action at a distance', and claimed it showed something was wrong with the theory. He was wrong, it turned out. That spooky action at a distance is real. And it makes quantum computers extremely powerful.

Put three particles into superposition and entangle them together and you simultaneously hold all the numbers from binary 000 to binary 111 (decimal 0 to 7). Any single operation you perform with those three particles, a sum or a multiplication for instance, is done on all eight possible numbers at once. Computing using quantum particles speeds things up considerably.

Take what it can do to the problem of factoring, for example.

Given any large enough number, it's difficult to find two numbers that, multiplied together, will create it. These numbers are its factors, but finding a large number's factors is a matter of trial and error. If the large number is 15, there's not a lot of trial to perform: you can quickly find the factors of 15 to be 3 and 5. But if the number looks like this:

74037563479561712828046796097429573142593188889231289084936232638972765034028266276891996419625117843995894330502127585370118968098286733173273108930900552505116877063299072396380786710086096962537934650563796359

it's a little more difficult. The codes that offer us secure banking, secure Internet transactions and secure telecommunications depend on this number being extremely difficult to factorise. The number above comes from a factoring challenge put out by the security firm RSA. It was solved in 2012; the factors are

90912135295978188784406583026004374858926083103283587204285121689604115286409333678249507883679567568061

and

8143859259110045265727809126284429335877899002167627883200914172429324360133004116702003240828777970252499

Getting this answer involved the equivalent of running 500 computers for a year. A single suitably powerful quantum computer would do it in a split second. We know that because, in 1994, a researcher called Peter Shor showed that a computer operating by the laws of quantum physics could cut through the factoring problem like a hot knife through butter. The spectre of this new, powerful kind of computer was immediately worrying because our best cryptographic codes – the kinds of codes that are used to protect military secrets and financial transactions are reliant on factoring being hard to do. The United States National Security Agency (NSA) immediately poured enormous amounts of money into research looking at how a quantum computer might be built – and how fast that might happen.

Ten years later there was still no quantum computer. The NSA's panic abated, but everyone with an interest in maintaining standards of security was still keeping an eye on the field. What no one had yet realised is that a hugely powerful quantum computer already exists. You're living in it.

Even for a quantum computer, the universe is pretty impres-sive. Its computations can be carried out by various means. Atoms bouncing around are capable of carrying and processing information, for instance. A single atom can carry something like twenty binary digits (bits) of information and, as Fredkin had pointed out decades previously, two atoms can collide with an outcome that is entirely equivalent to the information processing that goes on within a computer. The concentration of various chemicals within a mix can also store bits: cause these chemicals to react together, and they too can process the information like a computer. Viewed from this perspective, the whole universe is busy performing computations.

Seth Lloyd has calculated just how powerful a computer the universe is. His calculation starts with something the size of a laptop computer. A litre of the universe – roughly a kilogram of matter – can perform a million billion billion billion billion operations every second.

That processing power is applied to about 10 thousand billion billion billion bits of information. Since time began, Lloyd has calculated, the universe has performed around 10^{122} operations on 10^{92} binary digits. What are those operations? We see them as chemistry and physics, as the processes of life and the mechanisms of thought. Your actions are programming the universe. 'We are clay,' Lloyd says, 'but we are computational clay.'

To understand how human beings play a role in determining the program the universe is running, let's begin with a visit to Anton Zeilinger's labs in Boltzmanngasse – a street named after Ludwig Boltzmann – in Vienna. Zeilinger's group is most famous for its quantum interference experiments. The phenomenon of interference has been known for hundreds of years. When waves of any description – water waves, or the waves we associate with light – pass through a narrow aperture, they spread out as if emanating from that point. Put two apertures next to each other, shine a single light on them, and it's like having two lights rather than one. In the right set-up, these emanating waves will interfere with one another.

To see an interference effect, we have to match the size of the apertures, the distance between them and the wavelength of the light (that is, the distance the light travels to complete one cycle of its wave oscillations). If the set-up is right, the two light sources will have an effect on each other that creates a characteristic interference pattern. If you stand at the correct distance, facing the light sources like a prisoner facing a firing squad, then shuffle sideways, moving parallel to the apertures, you'll perceive alternating moments of brightness and dark. At certain points, the lights cancel each other out. Shuffle sideways a little more, and you'll find yourself brightly lit where the lights reinforce each other. Shuffle along again, and you'll pass through another zone of darkness before you encounter the next region of blinding illumination.

All this can be easily explained when the light is treated as a wave. Where the two sources are both at the peak of their cycles, things are bright. Where their troughs meet, things are truly dark.

Zeilinger's experiments put a little spin on this. The researchers turned down the intensity of the light source so that it emitted just one particle of light – one photon – at a time. Only one photon hit the narrowly spaced apertures at any one moment, and so there's just one photon, coming through just one aperture, at any moment. With just one photon there can be no interference effect, no pattern of light and dark regions. Or so you would be forgiven for thinking.

It's true that if you look to see which aperture the photon goes through, a perfectly reasonable thing to do, there is no interference pattern produced in the far distance. The strange thing is that, if there is no observation, the interference pattern returns. There is only one possible explanation for the return of the interference pattern. Since there is only one photon of light in the apparatus, the photon must have gone through both slits at once.

This is the source of the quantum computer's power. As we have seen, a quantum particle can be in two states – 'on' and 'off', or 'here' and 'there' – at the same time. What's interesting is that when someone or something tries to observe this strangeness, or someone ignorant of the strangeness tries to log which of its possible states the particle has adopted, the strangeness goes away. It's the question encapsulated in the title of David Lindley's excellent book on quantum theory, *Where does the Weirdness Go?* The answer to that question can be found by looking at the information whizzing around in the experiment – especially when Zeilinger's team do their experiments using balls of carbon molecules, known as fullerenes, in place of the photons.

Using fullerenes allows the researchers to add a new variable: temperature. It turns out that heating the fullerenes destroys the interference pattern as surely as trying to see which slit they go through. Why? Because hot molecules emit heat radiation, and that radiation carries information about the position of the molecule. Examine the radiation and you can tell, to a certain degree of accuracy, where the fullerene was. Which means you would be able to tell which slit it went through. Which forces it to be one or the other, and not both. Which means no interference pattern.

Different temperatures cause different amounts of information to be lost from the fullerenes, blurring the distinction between the dark and light areas of the interference pattern. What was once black and white becomes elephant grey.

Zeilinger and his team were able to calculate what kind of radiation would be emitted by their fullerene molecules at different temperatures, and from that they could determine how much information about the fullerene would be lost. Then from that they worked out how greyed-out the interference pattern would be at various different temperatures. Their predictions matched perfectly with their experiments.

What we can glean from this is that the photon, or the fullerene (or you or I, if we set the experiment up right), has no independent existence. It is not a wave, or a particle. It is neither and both, in a mysterious way that depends on its interactions with its environment. Looking at the experiment, or working at a certain temperature, brings a particle into existence. Other conditions make it exist as a wave. The exact nature of its existence depends on things other than itself, things to do with an exchange of information with the outside world. Things, in other words, to do with you and me, the programmers.

To see how we program the universe, let's re-imagine a thought experiment that John Wheeler first set out in 1978. It is called the 'delayed-choice experiment', and it involves light travelling to Earth from a distant star. Directly between the Earth and the star is an enormous galaxy, whose gravitational field bends the light's path, as Einstein predicted in his general theory of relativity.

The light can pass either side of the galaxy on its journey to Earth. According to quantum theory, that means a single photon of that light should take both paths – unless someone is watching which path it takes. But the distance between the galaxy and Earth is so large that we can make the choice of whether to watch or not long after the photon has passed the galaxy. If we do that, what happens? Will there be interference, or will there be a single photon? How does the photon know, before it is too late, whether its information is to be released to us?

The experiment is difficult to do over cosmic scales, but versions of it have been done in laboratories using equipment similar to Zeilinger's interferometer. These experiments show that the experimenter's choices can indeed influence the outcome of an experiment long after the result should already have been determined.

Clearly there are issues of time at play here, and many questions about the role of a mind that is conscious of observing the experiment (as we have seen, consciousness is a slippery issue). But whichever answer is right, we become participators in the processes of the universe, as Wheeler put it. 'Physics gives rise to observer-participancy, observer-participancy gives rise to information, information gives rise to physics.' We are in a paper–scissors–stone situation where we cannot find the logic to disentangle ourselves from the universe. It is the ultimate computer, and it gave rise to us. Now we bring its computations to life. 'Some part of our being knows this is where we came from,' Carl Sagan once said. 'We long to return. And we can. Because the cosmos is also within us. We're made of star-stuff. We are a way for the cosmos to know itself.'

A half-hour's drive south-east from Zeilinger's labs is Zentral-friedhof, Vienna's enormous cemetery. It houses the remains of Beethoven, Schubert, Brahms and Johann Strauss. It is also the last resting place of Ludwig Boltzmann. Grave 1 of group 14c, on the right as you approach the cemetery church of St Charles Borromeo, even has Boltzmann's most famous equation inscribed on the tombstone. $S = k \cdot \log W$ describes the entropy or disorder of a system; this is the S of the formula. W describes the various ways of arranging the atoms in the system; k is a fixed number now known as Boltzmann's constant. It is a way of showing that the second law of thermodynamics always holds.

Boltzmann didn't invent thermodynamics; it is a theory that preceded him and outlived him, surviving the onslaughts of scientists far more robustly. It began, in its modern form, as an attempt to create more efficient engines for powering the industrial revolution. A century later its worth was considered so great that the second law of thermodynamics – essentially, that the disorder of the universe is always constant or increasing – was elevated to a near-untouchable status. 'If your theory is found to be against the second law of thermodynamics I can give you no hope,' wrote astronomer Arthur Eddington in 1915. 'There is nothing for it but to collapse in deepest humiliation.'

After almost another century, the second law gained another victory: Vlatko Vedral and his colleagues analysed the uncertainty principle of quantum theory, which determines how much information can be extracted from a physical system. They found that it is essentially another way of writing the second law. The probabilities inherent in quantum theory link to the statistical laws that govern thermodynamics. Both link to information theory. In 2012 Vedral, Markus Müller and Oscar Dahlsten published a paper outlining how information theory, quantum theory and thermodynamics seem to be intertwined in a way that suggests a new theoretical framework for the universe is possible, one in which quantum theory, relativity, time and gravity all emerge as consequences of the physical nature of information. God, Vedral says, is a thermodynamicist.

Vedral and his colleagues are working at the edge of uncertainty here – it is impossible to say yet what they can and will achieve with this approach. Hopefully, then, we've still got time to fix the standard story of the universe before they take us back to before the beginning. That Big Bang we hear so much about is starting to get a little smaller, a little weaker. Fortunately, we might be able to fix it by taking the universe for a little spin.

9

COMPLICATING THE COSMOS

The story of creation is far from complete

Though a good deal is too strange to be believed, nothing is too strange to have happened.

Thomas Hardy

t is March 2013, and in the front row of the plush Geneva International Conference Centre, physicist Alan Guth is having trouble staying awake. He has flown in from Boston and nothing can keep his jetlag at bay, not even the sight of Oscar-winning actor Morgan Freeman pontificating from an autocue on the achievements of modern physics. Every few minutes Guth's wife leans in and wakes him with a gentle word. Then, a few seconds later, Guth falls asleep again. It's not as though he's needed. Guth has already been up on stage and collected his cheque for $3 million.

The money comes from Russian billionaire and entrepreneur Yuri Milner, who wants to reward the biggest thinkers in physics with a prize – the Fundamental Physics Prize – that aims to 'one-up' the true Oscars of science, the Nobel Prize. But it's hard to imagine Guth sleeping through the Nobel ceremony.

Guth received the Fundamental Physics Prize for the invention of inflationary cosmology. And though inflationary cosmology, which suggests that the universe went through a brief period of ultra-rapid expansion just after its birth, is a good idea; many physicists suggest it creates far more problems than it solves.

In 2008, at a week-long seminar in Cambridge, England, Guth ended up falling out with one of the theory's eminent detractors, Paul Steinhardt of Princeton's Institute for Advanced Study. After their disagreement, they didn't speak to one another for the rest of the week. Another delegate, Michael Turner of the University of Chicago, went on record describing the theory of inflation as 'duct-taped' and perhaps within a decade of falling apart. That's a worrying claim because Guth's theory is absolutely central to our best story of how the universe came to be as it is. But inflation is far from being the only weakness in the foundations of our cosmic history.

There is, for instance, a problem with the abundance of some of the elements – notably lithium. Atoms of this substance just don't exist in the amounts they should when working from what we think we know of how the universe began. There are certain structures in the universe – large masses of matter, and large regions of too-empty space – that undermine the standard cosmology. There are the now-familiar problems of dark energy and dark matter, stuff that seems to be out there but evades detection and may be hinting that we have fallen prey to a cosmic illusion. Then there's the discovery that everything – including the universe – is spinning. Even that stalwart of standard physics, the Higgs boson, is being difficult. The way it made its Nobel-winning appearance at the Large Hadron Collider at CERN in Geneva has lent support to a theory that does away with Guth's period of inflation. Every time we declare ourselves to be in a Golden Age of Cosmology, we are assailed by another phenomenon that cries, 'But what about this?' The Big Bang appears to be under attack from a swathe of Big Buts.

Perhaps the best place to start this story is with a young Russian boy called Georgii Gamov. Georgii grew up in Odessa, on the north-western shores of the Black Sea. The city is most famous as the site of the 1905 Potemkin uprising, where scores of the city's residents were massacred by Russian imperial troops. Five years after that pivotal moment in Russian history, six-year-old Georgii watched Halley's comet pass overhead from the roof of his parents' home. That moment, he later said, was when he fell in love with scientific enquiry.

In 1926, obsessed by the new quantum theory being developed in western Europe, he romanised the spelling of his name to Gamow – his aim was, it seems, to increase the chances that a paper he sent to the journal *Zeitschrift für Physik* would be accepted. Two years later, impressed by the promise of their student, his university sent him to study in Germany for four months. On the way home he stopped off in Copenhagen to call on Neils Bohr, the founding father of quantum theory. George impressed here too, and Bohr arranged for him to stay in Copenhagen for a month, and then in Cambridge, England, the following year.

George Gamow was, by now, infatuated with western Europe, a fact that was noted at home. Gamow repeatedly outstayed his designated return dates on scientific trips abroad. His public mocking of the poor state of Soviet science at every opportunity won him enemies. On more than one occasion the authorities confiscated his passport. There was an unexpected upside to this; it was on one of the many long days he had to spend at the passport office that he met his future wife, the physicist Rho Vokhminzeva.

Within a few years they had concocted a plan to flee the Soviet Union in an unauthorised defection. In the spring of 1932 Gamow and his wife set out to cross the Black Sea in a canoe; unfortunately a storm forced them to abandon the journey. That November, however, some smooth talking gave Mr and Mrs Gamow the opportunity to leave the Soviet Union. They never went back.

Gamow was also rather good at fleeing areas of physics that he felt he was done with. That's why he made such profound contributions to wide-ranging areas of the subject without ever becoming a household name. It was Gamow who came up with the theory of Big Bang nucleosynthesis, showing how elements would form in the fireball of the universe's first moments. Gamow also made major progress in stellar theory, showing what goes on inside fiercely burning stars. In 1948, in the same year that the celebrated scientists Thomas Gold, Hermann Bondi and Fred Hoyle proposed a mathematical model for the Steady State universe – a universe that had no beginning – Gamow and two colleagues set out to prove that there was a beginning. They examined what would happen if the universe began with a fireball of energy and matter and concluded that today's universe would be filled with photons of a particular, rather low temperature. Nobody knew how to look for this 'cosmic microwave background radiation', and the idea was quickly forgotten. Only in the mid 1960s did astronomers stumble across it.

Arno Penzias and Robert Wilson won the Nobel Prize for the discovery, the first proof of a Big Bang. Gamow's contribution to the Big Bang theory was forgotten for decades, but his consideration of the details is still largely unknown. However, in a few corners of physics, it is making a comeback. It's becoming clear, for instance, that we can no longer ignore the turbulence.

You might know the word 'turbulent' from your history lessons.
Henry II almost certainly never said, 'Will no one rid me of this
turbulent priest?' about the Archbishop of Canterbury Thomas
Becket, but it remains among the best-known historical quotes
learned by British schoolchildren. Turbulent, to historians,
means troublesome. To physicists, it also means trouble – but
only because they find it so difficult to deal with. Turbulence
occurs when things move in complex and unpredictable ways.
The Nobel Prize-winning physicist Richard Feynman once
described turbulence as 'the most important unsolved problem
of classical physics'. It is still unsolved; perhaps that's why no one
has ever wanted to face Gamow's idea.

In 1954 Gamow had moved on from the idea of cosmic radiation and started considering the moment when galaxies pulled
themselves together from the gas of particles that filled the early
universe. He thought it a sensible move to assume that 'the primordial gas was in a state of large-scale irregular motion, that is,
it was in a turbulent state.' It was a prescient assumption.

Turbulence is one of those things that you know when you
see it. A river flowing gently along, say the Thames flowing
through London, is definitely not turbulent. But the quick-mov-
ing white water of a mountain stream can indeed be turbulent
– in parts, at least. Turbulence requires, for the most part, flow
in three dimensions, the occasional appearance of vortices –
something like whirlpools – and unpredictability. If the flow is
turbulent, you can't guess where a particular molecule of water
will be in a few minutes, even if you know everything about the
forces acting on that molecule, as well as its position and velocity right now. It's turbulence that makes predicting the weather
so hard – turbulent flow in ocean and air currents mean we
can never be quite sure where cold and warmer air or water
will end up.

But it's the vortices that are the real hallmark of turbulent flow. These are almost like solid structures surrounded by flow – think of hurricanes, tornadoes, or, less terrifying, the smoke ring, where the blower makes invisible vortices visible by congregating smoke particles around the vortex's walls. On YouTube there is a wonderful video of an A340 airliner descending into fog: the wingtips of aircraft always create vortices, but here you can see them form; it's quite beautiful.

A closely related feature is the eddy current. Definitions are again loose here, especially defining the difference between an eddy and a vortex. An eddy tends to be small-scale curved currents created by the flow of a fluid after it has been forced past an obstacle in its path. A vortex forms more spontaneously, and is defined by flow in a complete circle.

Whatever the tight definition, both are likely to have existed at the moment of the Big Bang when the primordial gases flowed in conditions of enormous temperatures and pressures. It is these vortices and eddies, Gamow said, that seeded the structures we now see as galaxies.

There are some striking sentences in Gamow's paper, published in the *Proceedings of the National Academy of Sciences*. He says, for instance, that eddies in the gas would have created areas of high and low density, where the gas is compressed or stretched out. Because this gas was the trigger for the formation of galaxies, we should expect to find areas of high galaxy concentration – large clusters – and areas where there are far fewer, or almost none. The distribution of galaxies throughout the cosmos, he said, represents a 'fossilized turbulence'. Looking now at the way galaxies are distributed through space certainly provides evidence for this initial turbulent state. But at the time few people were interested.

It's probably worth noting that, at this point in scientific history, hardly anyone believed there was a Big Bang. To most minds, it made far more sense to consider the universe as existing from everlasting to everlasting in Gold, Bondi and Hoyle's unchanging Steady State. The one person who did look was astronomer Vera Rubin.

Rubin is better known as the woman who rediscovered dark matter, something we'll circle back to a little later. Her PhD thesis, though, followed up on Gamow's suggestion. It was an astronomical study of the motions of galaxies, and it showed that the patterns of galactic motion were a good fit to Gamow's models of turbulence. The thesis was published in 1954, and provided a rare moment of support for the Big Bang model.

Rubin's results were never published in a prominent journal and disappeared almost without trace. That could be because science is always reluctant to give up its favourite ideas – in this case, the Steady State universe. It might be because turbulence is, as we have already said, so difficult to handle. We still don't know how to solve the mathematical equations that model it, after all, which is why our equations concerning the evolution of the universe have simply left it out. If that seems negligent, it's also pragmatic: it would have been far more difficult to complete the Big Bang model if we tried to take turbulence into account. The trouble is, that pragmatism seems to be coming back to bite us.

A central pillar of our cherished Big Bang theory is the assump-tion that the universe is the same in all directions, that there is no up or down or left or right or clockwise or anticlockwise. It's called the 'Cosmological Principle' and it says, roughly, that the universe looks the same wherever and whoever you are, and whichever direction you're looking in. Our equations for how the universe evolved rely on this assumption. If it is wrong, every-thing falls apart. That's why everyone was so horrified when João Magueijo and Kate Land found the Axis of Evil.

It is typical of Magueijo to come up with such a mischievous name. He champions difficult causes, offers dissent as a matter of course and celebrates the outsider – all with a typically Por-tuguese arrogance and charm. Many researchers would have glossed over or ignored the anomaly in the data coming from the Wilkinson Microwave Anisotropy Probe (WMAP). Maybe many people did. But not Magueijo – no, he circled it with a big, fat highlighter pen. Then he published a paper focusing in on the 'anomalous alignment' and 'uncanny' correlations in the data. And he published it, no doubt with a smirk, under a name that echoed a politician's favourite term for a global terrorist network. Because this was cosmic terror.

Since its launch on 30 June 2001 the WMAP satellite has been gathering information on the cosmic microwave background radiation, the sea of photons that fills the universe, just as Gamow predicted in 1948. The photons were created just 370,000 years after the Big Bang, and their properties give us clues to the state of the universe at that time. It is from this radiation, and a few other angles of enquiry, that we deduced our current estimates of the age and composition of the universe.

The microwave background photons are all of roughly the same temperature: just below 3 Kelvin, or –270 °C (Gamow's 1948 paper suggested 5 K – not far out). However, WMAP exposes subtle differences in their temperature. Most of these deviations are randomly distributed throughout the cosmos, sup-porting the idea that the universe is the same wherever you are in the universe, and in whatever direction you choose to look. But Magueijo and Land spotted a region where certain of the devia-tions seemed to line up.

At first everyone said Magueijo and Land were wrong. Then they said it must be some kind of illusion, one of those statistical flukes that sometimes lead scientists astray. By 2010 that was the WMAP team's official line: it's there, but it's nothing to worry about. The trouble is, it was still there when the Planck telescope, the follow-up to WMAP, looked. Planck has better resolution than WMAP, and is ten times more sensitive. With the Planck data, the Axis of Evil was suddenly causing a real problem to the Cosmological Principle. At this point, though, it wasn't the only problem.

Between the constellations of Centaurus and Vela there is a largely blank patch of sky that, in 2008, garnered a great deal of attention. A team of astronomers spotted an enormous cluster of galaxies racing towards this empty space. The galaxies were moving at close to 100 kilometres per second, causing cosmologists quite a headache. As we have said, the universe is not meant to contain special structures doing something different to the rest.

The man who found this anomaly, now known as Dark Flow, is called Sasha Kashlinsky. It wasn't an accident, exactly, but Kashlinsky's quest does seem, in retrospect, a little quixotic. He was questioning the assumptions underlying our model of the universe, deliberately searching for things that everybody said really shouldn't exist. He had narrowed this down to one target: super-large clusters of galaxies. He was, in cosmological terms, a unicorn hunter. And to everyone's surprise, he found a unicorn.

The only way to detect giant galaxy clusters is to examine the photons of the microwave background radiation. The photons' energy is subtly altered by interactions with seas of hot gas in which the superclusters sit. But the changes are so subtle that only the biggest clusters of galaxies have a noticeable effect.

An even more subtle effect is imprinted on the photons by the movement of the clusters. A bounce off moving hot gases will change the photons' energy in a Doppler shift, the same effect that changes the pitch of an ambulance siren as it passes by. For some reason Kashlinsky thought it was worth looking for the Doppler shift in microwave background photons that had encountered large clusters of galaxies. Diehard fans of the Big Bang theory might wish he hadn't.

Galaxies, galaxy clusters and everything else in the universe should be ambling around in unremarkable ways, not speeding through the cosmos as if late for an appointment. That's why Kashlinsky and his team didn't tell anyone what they had found for a whole year: they didn't want to look stupid. So they checked their data. Then they checked it again. When they finally published their tale of galaxy clusters streaming through the universe at breakneck speed, all in one direction, they got the inevitable response: you've done something wrong, and further analysis will make it go away. But the Dark Flow hasn't gone away. It's got worse.

Kashlinsky initially found 800 galaxy clusters streaming towards the gap between Vela and Centaurus. Looking again, they found 1,400 of them. The stream of clusters speeding away from us can be seen as far away as 3 billion light years from Earth. It's as though they're racing to escape the universe.

Many people have argued that the clusters must be experiencing a gravitational pull from some enormous structure just beyond the edge of the visible universe (we can only see as far as the speed of light, combined with the age of the universe and its expansion rate, will allow). That in itself is problematic for the Big Bang theory because we don't have anything that big in our part of the universe, and – remember – things are meant to be the same everywhere.

They're clearly not. Since we spotted the Dark Flow, we have also found a supergiant structure. At 4 billion light years long, the Huge-Large Quasar Group (physicists are terrible at giving things names), spans one-twentieth of the universe's diameter. These are the kinds of distances over which you're not meant to see anything remarkable. It's another unicorn in the hallway of cosmology.

Hunting astronomical unicorns is a thankless task: the discovery has left Kashlinsky branded as 'controversial', and he has become embroiled in plenty of heated arguments about his work. It's a situation that John Webb knows well.

We've had the Axis of Evil and the Dark Flow. Now, if you're ready to face even more chaos, we can contemplate the Axis of Alpha: a line that suggests the laws of physics are different on different sides of the universe.

Alpha is one of the fundamental constants of physics. It is a number, roughly equivalent to 1/137, that gets plugged into the equations that determine how light and matter interact. Those equations are important in many branches of physics: they determine things as disparate as the colour of paint and the energy contained in empty space, for example. However, it seems that this constant, a central pillar of physics, is anything but constant.

John Webb is the astronomer who spotted this problem. Initially Webb used the Keck telescope on Mauna Kea in Hawaii to study the light coming from distant quasars. He determined that when that light was passing through interstellar gas clouds on its way to the Keck, alpha was slightly different from the value it has when we measure it on Earth.

When light interacts with a cloud of gas, a portion of the light's energy gets absorbed by the electrons in the atoms (or

molecules) of gas. That absorbed energy corresponds to a particular frequency in the spectrum of light, and the exact figures
involved depend on alpha. What Webb noticed when the quasar
light passed through the interstellar gas clouds was that the wrong
bits of light were absorbed. Some were at higher frequency than
you would expect for the same interplay between light and matter
on Earth. Some were lower. Every way Webb tried to cut it, the
only explanation that fitted the evidence was that in that gas–
light interaction all those billions of light years away (or billions
of years ago, depending on how you want to look at it) alpha was
around one part in a million smaller than the value we cherish
today. However, Webb has now found evidence suggesting that,
instead of changing over time, the laws of physics might change
over space. There are, in effect, local by-laws that subtly change
depending on your location.

Webb is not working alone on this. A whole slew of researchers are sharing his burden. Michael Murphy and Julian King, for
example. It was King's job to check the Keck results, gathered
from northern hemisphere skies against the same kind of data
gathered at the Very Large Telescope (VLT; I said physicists were
hopeless with names) that surveys southern skies.

Viewed from the other side of the Earth, alpha was not smaller
but bigger than we are used to. Imagine extending a line from pole
to pole out into the universe. The further 'south' you go through
the cosmos, the larger the value of alpha appears to be. As you go
'north', on the other hand, it gets smaller. Murphy summed it up
beautifully: 'If we didn't violate the laws of physics in our previous
results, we're certainly violating them now.' What's more, Webb's
Axis of Alpha seems to line up with the direction of Kashlinsky's
Dark Flow. That makes it even harder to dismiss as, to use NASA's
pronouncement on the Axis of Evil, a statistical fluke.

If Webb is right, the Axis of Alpha ruins all our laws of physics, as well as the history of the universe they have helped us create. Alpha may vary in time and space, meaning that the laws of physics could have been different at the Big Bang.

Webb's analysis is ongoing. Many people refuse to believe it can be correct, and it will take some almighty statistical power in Webb's evidence to bring the doubters round. But, at the moment, no one has managed to find the flaw in Webb's work. He has been doing it for nearly twenty years without serious challenge to his methodology, his data or even the conclusions he draws. All he ever hears is, 'I don't believe it.' And, 'Oh no, not John Webb again.'

He doesn't hear that from Michael Longo, though – Longo thinks all these lines might be explained by his observation that the universe has a spin.

'Such [a] claim, if proven true, would have a profound impact on cosmology and would very likely result in a Nobel prize.' The quote comes from a reviewer's report on a paper by Michael Longo. It's not clear whether the reviewer was scoffing or impressed. But Longo's claim that the universe has a net spin – something that Gamow would no doubt take as obvious – has to be taken seriously.

Longo and his team used a telescope at Apache Point Observatory in New Mexico to look up at the sky and note which way each of hundreds of thousands of galaxies are spinning. By studying the curves of their spiral arms, you can infer the direction of rotation. In most of the sky that direction appears random. But you can draw an imaginary line in the sky that sits 10 degrees off the axis of our own galaxy's spin. Look along this line and you'll find more galaxies spinning one way than the other. Out of 15,000 visible galaxies, roughly 7 per cent more are 'left-handed' than 'right-handed'. The chances of this being a statistical fluke are around 1 in a million. Especially since, when you look at the southern sky, you see the same effect, but in reverse: more right-handed spirals than left.

The galaxies – which, let's remember, Gamow and Rubin showed are most likely the end result of turbulent flow of matter just after the Big Bang – are rotating like bicycle wheels. The result of all these spins is a net angular momentum to the universe. If there's one thing physicists know, it's that angular momentum is like energy: it can't be created or destroyed – it can only be transferred from one kind of carrier to another. That means the universe was born with a spin. And if it has a spin, it also has an axis. And if there's one thing a universe isn't meant to have, according to the standard story, it's an axis.

We are done with the axes that cause such problems for our standard view of the Big Bang story, but we are not done with the problems. Before we finish, we must look at two more things that undermine the neat tale cosmologists have constructed. The first is to do with lithium.

You are probably most familiar with lithium because it powers your mobile phone and laptop – or, after a few hours' use, doesn't. But next time you shake your head at your dead lithium battery, take a moment to get some perspective. As we will learn in a later chapter, you have a very blinkered concept of time. The atoms in your body were forged in the explosions of supernovae many hundreds of millions of years after the Big Bang. That makes them cosmic youngsters compared to the lithium atoms so vital to your mobile phone battery. These atoms were created in the first three minutes of the universe's life.

There is something extraordinary about holding something so

old in your hand. But lithium is as much a source of concern, as it is of wonder. What we know of the abundance of lithium in the universe doesn't tally with the chemical legacy that should have come from the Big Bang.

Here's the story so far. In the beginning there was energy. Some of that energy coalesced into what we call the fundamental particles: quarks and electrons. Triplets of quarks came together, via what's known as the 'strong force', to form protons and neutrons. The electromagnetic force pulled a proton and an electron together to form the first hydrogen atom. Hydrogen atoms fused and coupled with neutrons to form helium atoms. Eventually bigger atoms, right up to lithium, formed. Until the first stars lit up and eventually created supernovae with the necessary conditions to create heavier elements, that was it.

This period of activity is known as Big Bang nucleosynthesis (Gamow also pioneered the study of these processes). Researching it has been highly productive on the whole. The observed amounts of helium and hydrogen in the cosmos tally so well with what the theory predicts that this has become the most convincing evidence in support of the Big Bang. But this is where two French astronomers spotted a problem.

In March 1981, long before the telescopes on Mauna Kea gave us our first glimpse of the Axis of Alpha, they played host to another disruptor of the accepted story of the universe. François and Monique Spite were carrying out observations using the Canada–France–Hawaii telescope. They were trying to estimate the quantities of various elements remaining in a star contained within the constellation Hydra.

Since we know what colours the elements emit when they are plied with energy, the light from a burning ball of gas can tell us what elements are there. Spread out the starlight into its various colours, in the same way that a water droplet spreads sunlight into a rainbow, and you can infer what's creating it. The results the Spites inferred from HD 76932 pointed to a big problem with lithium.

HD 76932 is twice as old as our sun. Its lithium was depleted – burned up in thermonuclear reactions. But not as much as it should have been. The chain of reasoning is complicated, and beyond our scope here, but the abundance of lithium in this old star rang alarm bells. The Spites published a paper the following year that contained an exhaustive analysis of the implications. The main one was that the universe should have more lithium than has ever been accounted for.

More than three decades later, the anomaly remains. In fact, it's got worse. The Spites were looking at lithium-7, an isotope of the element whose atomic nucleus contains three protons and four neutrons. We now know that the cosmos contains one-third the amount of lithium-7 that the Big Bang theory says it should. We also know now that there is too much lithium-6, which contains one less neutron in its nucleus. One thousand times too much, to be precise.

So far we have a problematic element that doesn't fit the story, added to our four violations of the Cosmological Principle. Oxford University's Subir Sarkar summed up the discomfort such discoveries engender in science. Accounting for them will make cosmology 'too bloody complicated', he told *New Scientist* magazine. But, thanks to inflation, dark energy and dark matter, it's already more complicated than anyone would like.

Let's start with the first 'dark universe' discovery. It was way back in 1933 that a Swiss astronomer called Fritz Zwicky thought something must be missing. He noticed that clusters of galaxies were spinning at speeds that meant they should be ripping themselves apart: the centrifugal forces ought to be big enough to overcome the gravitational attraction the stars had for each other. He worked out that there must be more matter out there than could be seen – the gravitational pull of this 'dark matter' had to be holding the galaxy clusters in place.

Zwicky's idea was largely ignored; it was only when Vera Rubin came to similar conclusions in the 1970s that people took it seriously and began to look for the dark matter. Astronomers catalogue the contents of the universe in terms of its mass-energy, since mass and energy are interconvertible, as shown by Einstein's famous $E=mc^2$ equation (E is energy, m is mass and c is the speed of light). Our calculations suggest that dark matter makes up nearly a quarter of the universe in those terms – which makes it all the more astounding that we have yet to find any.

Neither, as it happens, have we found the source of the 'dark energy'. All we know about it is that it seems to be making the expansion of the universe speed up rather than slow down, as we would expect in a cosmos where everything is exerting a gravitational pull on everything else. Some have suggested that dark energy, whose existence is inferred from observations of the way certain supernovae brighten and fade in the sky, is another cosmic illusion that will disappear when we can abandon some of our assumptions about the nature of space and time.

For now the mainstream belief is that dark energy constitutes a smidgen under three-quarters of the universe. Dark matter being just under a quarter means that we have seen only a very small percentage – 4 per cent, roughly – of what's out there. Almost all of the standard-story universe is missing. Alan Guth's inflation is central to this story – hence the cheque for $3 million. But it, too, has its difficulties.

Inflation was meant to solve two problems cosmologists had when they described the story of the universe. They are called the horizon problem and the flatness problem.

The flatness problem is that the density of matter in the universe is uncannily 'just right'. There is just enough matter for gravity to stop the post-Big Bang universe's expansion running away uncontrolled, causing the cosmos to expand so much that no stars and galaxies could form. At the same time there is not so much matter that the universe immediately collapsed back in on itself. The fine-tuning required for this will astonish you – the density of matter in space is within 1 part in 10^{57} of the required value. It certainly astonishes cosmologists, who are naturally suspicious of coincidences like this.

The horizon problem is fairly simple to describe, if not to solve. It is this: the universe is incredibly boring. Its diameter is something like 26 billion light years, and it all looks pretty much the same. It is all at roughly the same temperature, as evidenced by the incredible uniformity of the cosmic microwave background radiation. The problem is, in the normal scheme of things, it shouldn't have had time to become so bland.

As Gamow suggested, the beginning of the universe was a raging furnace of energy, but not everything would have been at exactly the same temperature. Quantum theory requires that there were fluctuations in energy that would have evolved into hot and cold spots in the first moments of the universe. Because hot spots lose heat by radiating into colder areas, we expect that everything eventually reaches the same temperature. However, that radiation is the transfer of photons, which travel at a limited speed: the speed of light. Looking at the universe as it is today, there simply hasn't been time for photons to travel between all the various areas of the cosmos – a necessary part of making the temperature as even as it is. Put simply, the universe is too big to be at such an even temperature.

Guth's solution to the horizon problem was to suggest that the universe evened out its temperature when everything was in close proximity. After that was done, he suggested, the universe suddenly blew up in size, inflating like a balloon being filled by the lungs of Hercules. The figures are staggering: in order to get to where we are today, this moment of inflation blew up the universe so that it was 10^{60} times bigger in just 10^{-35} seconds. Those figures are the ones used by the NASA Planck mission; if that seems intangible, try these from the general NASA archive: during the inflationary period, the universe 'grew from a subatomic size to a golf-ball size almost instantaneously.' It's not quite as precise, but, given the uncertainty over inflation, it's probably less wrong than the Planck version of events.

According to the standard inflationary cosmology, the rapid blowing-up of the universe stopped almost as soon as it had started. It's almost like a marble rolling off the top of a hill: suddenly, it's moving very fast, then, when it reaches the bottom of the hill it slows down dramatically. Since the end of inflation, the universe's expansion has been more pedestrian.

As well as telling us how the universe got its even temperature, inflation also solves the flatness problem by spreading out matter and flattening the curves in space that a high-matter density creates. It's enough of an achievement to put $3 million in Guth's bank account. However, in reality it only pushes the question a little further away. The obvious question to ask is how this could happen – and we really don't have an answer for that.

Inflation is not just one theory. It is an umbrella term that describes a plethora of ways – or models – in which the universe might blow up. In some it blows up at a constant pace. In others it speeds up or slows down, or follows complicated patterns of speeding and slowing. There are hundreds of possibilities, which means physicists have to try to disprove each one of the models by looking for clues to what actually happened during inflation.

That was one of the tasks of the Planck mission. However, when it reported its results in March 2013, they put inflation in some serious difficulties. As Princeton theorist Paul Steinhardt told *Nature* reporter Zeeya Merali, 'if you take the data we've been given and just follow your nose, then inflation and the whole Big Bang paradigm seem to be in big trouble.'

To be fair, Steinhardt didn't really want to get involved with this. He has been arguing for years that inflation is a conjuring trick, an embarrassment. No one has really taken much notice, and he has become quieter in recent times. He was willing to let the whole Planck thing slide. But then a young Harvard astronomer–philosopher called Anna Ijjas burst into his office and demanded to know whether he was going to let people get away with telling such outrageous lies about inflation.

Ijjas held in her hand a paper written by the team running the Planck telescope mission. It said that 'the simplest inflationary models have passed an exacting test with the Planck data.' Ijjas was outraged. The details of the paper showed that the Planck data did precisely the opposite.

Planck's observations suggest that the simplest inflation models – for example, the ones where it starts, proceeds at a uniform speed and then stops – are ruled out. All that is left are the more complex – Ijjas would say more unlikely – models known as 'plateau models'. In these models inflation starts very slowly before gathering speed, as if that hill has an almost flat top, so that the marble takes a lot of time to get going. The problem is, to get sufficient inflation from these requires that the universe be rather bland and uniform to begin with. That's exactly the wrong starting point; it's meant to be hot and turbulent. Bland and uniform is the end point.

Eventually, Ijjas persuaded the jaded Steinhardt, and Harvard's head of astronomy, Avi Loeb, to join her in a quest to make this problem public. The result was a paper called 'Inflationary Paradigm in Trouble after Planck 2013'. The Planck mission has ruled out all but a handful of possibilities for inflation, the paper said. Worse, the ones that disappeared were far more 'natural' candidates than the inflation models that provide the best fit to the cosmological data, making them even less likely to be useful resolutions to the horizon problem and the flatness problem.

What's more, those pesky Higgs hunters at CERN in Switzerland have made things even worse. Which is ironic because they, too, were sitting in the front row at the Geneva International Convention Centre, picking up their own cheque for $3 million. At the time of the ceremony in Geneva, only a few people realised there was a conflict between the Higgs discovery and Alan Guth's ideas on inflation. The conflict arose from something seemingly innocuous: the boson's precise location.

As any visitor will tell you, the CERN campus, situated on the outskirts of Geneva, is notoriously difficult to navigate. Its building numbers are placed largely at random: without a map, all you can do is wander and hope you come across the building you are looking for. It was largely the same for the CERN researchers charged with finding the Higgs boson. In 1964, the same year the astronomers stumbled across Gamow's cosmic microwave background radiation, a group of theorists said the boson ought to exist, but that there was no easy way to find it.

Physicists identify the location of particles by their energy. That was why, as the technology became available, accelerator physicists started searching for the Higgs boson by smashing particles into one another at various energies. When it didn't turn up in collisions of one energy, they tried another.

By 2001 we knew that the Higgs had an energy above 115

gigaelectronvolts (GeV). Another three years of work pinned it down to somewhere between 117 and 251 GeV. A few rethinks later, and it was placed, in 2012, somewhere between 115 and 152 GeV. The final answer was that the Higgs was found at 125.3 GeV. And that, oddly enough, is bad news for the Big Bang.

The Higgs boson comes from the Higgs field, which is supposed to fill all of space. It is this field that gives most of the universe's particles a mass: a resistance to acceleration or deceleration, and a response to gravity. Inflation is also supposed to result from a field – the inflaton field – that provides energy for the universe's sudden growth.

What we know about the Higgs boson pretty much rules out any inflationary models that were left standing after the Planck results. The interaction between the kind of Higgs field the newly discovered boson creates and the only inflaton fields left after Planck's data would cut inflation short. Stated bluntly, the universe would never have formed.

Steinhardt has a nice way of putting it. Instead of thinking of a marble on a hilltop, think about a marble dropping out of the sky above the Matterhorn. The likelihood of inflation having happened in a universe containing the kind of Higgs we have found, he says, is similar to the chances of that marble landing in a little dimple at the top of the pointy peak and staying there, rather than crashing down its steep slopes. Inconceivably unlikely, in other words.

It's only fair to say that Steinhardt has never been a fan of inflation. He has an alternative model, constructed with Neil Turok of the Perimeter Institute in Ontario, Canada. This is a self-recycling universe that doesn't have any period of inflation. It is not a popular model – certainly not as popular as the Big Bang plus inflation plus dark matter plus dark energy model.

Big Bang plus inflation plus dark matter plus dark energy: that probably sounds a little cumbersome to your ears. Be thankful, then, that there are no ad hoc fixes yet for the lithium problem, or the various axes that suggest the universe has a spin that might have arisen from Gamow's primordial turbulence. Once we start adding in solutions to those problems – if we can come up with them – the Big Bang theory will start to look less like a coherent narrative and more like a dreamscape: a mad whirl of disconnected stories. You might say, 'it is as though an artist were to gather the hands, feet, head and other members for his images from diverse models, each part excellently drawn, but not related to a single body, and since they in no way match each other, the result would be monster rather than man.' The quote actually comes from Nicolaus Copernicus – he is complaining about astronomers trying to fit their observations into the Earth-centred model of the solar system. Copernicus's observation, says philosopher of science Thomas Kuhn, is typical of those who see a scientific revolution coming over the horizon. Most scientists, Kuhn says, never see it coming – they are just swept away by the wave. Maybe it's time for Guth and the rest to wake up and smell the paradigm shift.

We'll leave it there for now – after all, it may be that we simply can't understand the true nature of the universe with the tools we have at our disposal. Perhaps we need another kind of computer, the one that Alan Turing dreamed up in 1936. We've had his basic machine running for decades, but Turing's 'hypercomputer' is something special. It can literally do the impossible – which makes it all the more exciting that, finally, it's under construction.

10

HYPERCOMPUTING

Alan Turing had another good idea

The imagination of nature is far, far greater than the imagination of man

Richard Feynman

The New York State Department of Health publishes a handy wall chart about chemical terrorism. One piece of advice concerns how to tell if you have been subjected to a cyanide attack. Your skin, it says, might go cherry-red. You may get frost-bite from contact with liquid cyanogens. You might suffer confusion, nausea or feel a desperate need to gasp for air. Nowhere, despite what you're probably thinking, does it mention the smell of bitter almonds.

Rare is the fictional detective who hasn't at some point talked about a smell of almonds and traced the cause of death to a dose of cyanide. The fact is, though, it should be rarer. Around 18 to 20 per cent of men cannot smell cyanide because they don't have the gene that is necessary to decode the molecule's properties. Only 5 to 10 per cent of women have the same deficiency. If someone is poisoning you with cyanide, you stand a much better chance of detecting their evil intent when you are a woman. Perhaps that was Alan Turing's problem.

After Turing's death in 1954, the coroner recorded a verdict of suicide. It certainly appeared, at first glance at least, that the mathematician had deliberately eaten a cyanide-laced apple. There are other possibilities, though. Oxford University computer scientist Jack Copeland laid them all out in an essay published in 2012. 'The evidence for suicide is very slim,' he says. Turing could have been murdered by the British secret service. The Cold War was escalating and Turing's sexual behaviour meant he was classified as a security risk and kept under police surveillance. On the other hand, he might simply have been careless. Turing kept cyanide in what he called his 'nightmare room', the small laboratory that adjoined his bedroom. He was known to be rather slapdash in his use of chemicals, and may well have unknowingly spilled some. If he was unable to detect the smell, his poisoning could have been through inhalation of cyanide gas, or the ingestion of cyanide left on his fingers as he ate that fateful apple.

According to Turing's biographer Andrew Hodges, suicide is almost certainly the right verdict – but with an added twist. In Hodges's view Turing introduced just enough ambiguity into his suicide for his beloved mother to believe it was the result of a clumsy chemistry experiment gone wrong. It is 'more credible', Hodges says, 'that he had successfully contrived his death to allow her alone to believe this.'

Copeland's essay, too, is inconclusive. 'The exact circumstances of Turing's death may always remain unclear,' he writes. 'It should not be stated that he committed suicide – because we simply do not know. Perhaps we should just shrug our shoulders, agree that the jury is out, and focus on Turing's life and extraordinary work.' It's easier said than done, because, as Copeland knows all too well, even Turing's work has been swathed in a swirling mist of myths and falsehoods. Especially his conception of a hypercomputer.

Turing is best known for his achievements in decrypting the communications transmitted using the Germans' fearsome Enigma machine. Enigma's encryptions, which profited from the machine's billions of possible plug and rotor settings, were thought to be uncrackable. Even before the outbreak of war, Polish cryptographers were trying to break Enigma, and had designed a *bomba* machine that cranked through the possibilities. When the invasion of Poland was imminent, the Poles shared their work with the French and the British – it was the *bomba* blueprint that inspired Turing to come up with his own 'bombe'.

The machine was built at Bletchley Park, the home of the Government Code and Cypher School where Turing spent the wartime years. After numerous people had lost their lives to create clues from which it could make deductions, the bombe did eventually break Enigma, and took two years off the war in Europe.

Turing's contributions to computing were far from over, however. In the years after the war, he took up residence in Richmond, where he designed a new computer. It was called ACE – the Automatic Computing Engine. During its construction, which took five years, Turing quit. His bosses at the UK's National Physical Laboratory frustrated him by insisting on certain design changes. Though the changes would suit the applications his bosses had in mind, Turing had other ideas. In 1948 he took up an offer to move to Manchester University, where enterprising scientists had already realised Turing's most prescient vision: the 'Baby'.

The Baby was a revolutionary design that stored its own programs. A small-scale experimental machine, it was quickly supplanted by the next generation: the more powerful Manchester

Mark 1. Together these marked the first steps towards the machines now in use around the world every second of every day and night. The workings of every single one of them, reduced to bare bones, follow exactly the same routines laid out in Turing's original design. And that, you may be surprised to learn, was based on the routines of a human mathematician.

Your computer – it is a fairly safe presumption that you own one, and you may even be using it to read this chapter – is not intelligent. It is, basically, a very, very capable, rigorous and efficient rendering of human capabilities. That is where the term comes from: the first 'computers' were human beings tasked with performing computations. In 1936 Turing imagined taking this to extremes and replacing a human with a machine that followed the same set of rules in order to get a computation done. As he wrote later, in the programmer's handbook for the Mark II Manchester computer, 'Electronic computers are intended to carry out any definite rule-of-thumb process which could have been done by a human operator working in a disciplined but unintelligent manner.'

The most generalised version of the computing machine is now widely known as the 'Universal Turing Machine'. A human being, given pencils and paper, infinite patience and attention and precise instructions, can do certain things. And a Universal Turing Machine can do them too.

Turing conceived his machine as composed of two main parts. The first is an infinitely long tape, divided into a series of squares as if it were a strip of graph paper. The tape can move left and right to allow the machine to read from, or write on, the squares; each square holds one symbol taken from a range of possibilities. The instructions in the machine tell it what to do on reading a particular symbol – it might be something like, 'replace the symbol with X, move tape one place to the right, then read the next symbol and follow the same routine'. If there is no applicable instruction, the computation halts, and the output is the series of symbols written on the tape.

It's rather hard to imagine this as being the root of everything your home computer does, but Turing showed that any set of instructions can be reduced to a mathematical operation that can be computed using such a machine – or by a human, of course. You can prove to yourself that you are a computer that does things by moving symbols around. Just set yourself a long multiplication problem. You will follow an algorithm you learned in school – one of many that mathematicians have created. All involve considering the position of digits, and writing them in specific places, just like Turing's universal computer.

Try 43 x 28. You might write the two numbers above each other, then write down the result of 8 x 3, then 8 x 4 to its left, carrying certain digits to the next position to the left. Then you'll move down a line, and across one digit to the left and do 2 x 3 and 2 x 4. Then you'll add the two numbers you obtained to get the final answer.

That's how I learned it. Maybe you do it differently. You could just multiply 40 by 20, then 40 by 8, then 3 by 20, 3 by 8, and so on. The results all get put into position so they can be summed in columns of units and tens, with some carrying over of numbers into other positions, to give the final answer. Another method, popular in the centuries before the printing press was invented, involves creating a lattice of four squares, then writing 43 across the top and 28 down the side. Then you divide the four windows diagonally, multiply each digit in sequence, putting the answer in the windows of the lattice, tens above the diagonal divide and units below it. Then you perform an addition through the diagonal lines to get the result.

This method fell out of favour a few centuries ago for purely technological reasons. The invention of the printing press was a great tool for education, but early presses couldn't print anything as complex as a matrix of numbers, and another, more easily printable, set of rules took over. But that's all it is: a set of rules. Whichever method you choose, you are recognising the relative position of the numbers involved, performing an appropriate multiplication (essentially a repeated summing operation) and then writing the result down in the right place in order to perform another operation on it at a later point. The electronic computer can do all of the above and much more. There is, however, one thing it cannot do – and this is the key to the power of the hypercomputer.

In April 1995, Microsoft boss Bill Gates was taking part in a live broadcast demonstration of what the forthcoming Windows 98 operating system would be able to do. His co-presenter plugged a scanner into the computer running the new system and announced that it would automatically detect the scanner and download the necessary drivers. Instead, to the horror of both men, the machine froze and the dreaded 'Blue Screen of Death' appeared on the display.

This is Microsoft's disaster alert. It means the computer will not go on. It means the machine has to be rebooted. 'I guess that's why we're not shipping Windows 98 yet,' was Gates's lightning-quick quip to the audience.

Though unfortunate for its inclusion in a live broadcast, such catastrophic software failures are hardly rare. Modern software is so complex that it is almost impossible to write it without making any errors. Microsoft's Word program, for example, is tens of millions of lines of code long, and no one person understands how the whole thing works.

You would think, given all this complexity, that there would be some automatic code-checker, something that can make sure there are no inputs that lock the machine up in the never-ending series of operations known to programmers as the infinite loop. However, that would require a program that could tell if a particular program ever reached an end point, or simply continued forever. And that, as Alan Turing showed in the years before the Second World War, is impossible for anything that can be represented by a Universal Turing Machine.

In 1931, a few years before Turing created the concept of his universal machine, the Austrian logician Kurt Gödel took the wind out of the sails of mathematics. He showed, in a paper called 'On Formally Undecidable Propositions in *Principia Mathematica* and Related Systems I', that every mathematical procedure is always based on something that is not provably true. There is, according to what came to be known as 'Gödel's incompleteness theorem', nothing that is entirely trustworthy about mathematics.

It came as a terrible shock to mathematicians around the world. Bertrand Russell's reaction was to declare himself puzzled: 'It made me glad that I was no longer working at mathematical logic,' he said. 'Does this apply to school-boys' arithmetic, and, if so, can we believe anything that we were taught in youth? Are we to think that 2 + 2 is not 4, but 4.001?' Someone rather cruelly compared Russell's efforts to understand Gödel to a dog staring at a blank screen.

Turing quickly grasped the Gödel incompleteness theorem, though, and expanded it to create the 'halting problem'. Just as it is impossible to say that 2 plus 2 definitely equals 4, rather than 4.001, it is impossible to tell in advance whether a program will eventually halt and produce an output, or will just continue running forever.

The proof of the halting problem harnesses Turing's knack for paradox. According to cryptographer Jack Good, who worked with Turing at Bletchley Park, Turing's stroke of genius in designing the bombe was in his use of a logic theorem that invoked the power of contradiction. In the right circumstances, everything you want to know is revealed by the machine when two apparent truths stand in conflict with one another.

The halting problem comes from a similar place. Here's a contradiction: 'This sentence is false.' If the sentence is true, then it is also false. Turing showed that you can feed a Turing machine a similarly paradoxical set of instructions to determine whether the machine will halt. The thing is, if the machine *will* indeed halt, those instructions make the machine loop endlessly. In other words, a Turing machine just can't handle that task. And that, in the end, is why you can't get the bugs out of code except by trial and error, or by going through the program line by line for every conceivable input – by running the program, in other words, albeit in your head.

The realisation that there is no way for a Turing machine to predict whether a program will finish running has had far more profound results than encouraging a fatalistic acceptance of buggy programs. It has also set false limits; people have either assumed a Turing machine can do everything, or assumed that the set of tasks that it is possible for any machine to perform is defined by what a Turing machine can do. Looking at it another way, people assumed that anything a Turing machine can't do is impossible.

Copeland has compiled a rogue's gallery of researchers who have misinterpreted Turing's thesis in this way. According to neuroscientists and philosophers Paul and Patricia Churchland, for example, Turing's results 'entail something remarkable, namely that a standard digital computer, given only the right program, a large enough memory and sufficient time, can compute any rule-governed input-output function.' The truth is, as we have seen, there are rule-governed input–output functions the Turing machine can't compute. That doesn't mean no machine could compute them.

Computer scientist Christopher Langton made the other mistake when he said that there are 'certain behaviours that are "uncomputable" – behaviours for which *no* formal specification can be given for a machine that will exhibit that behaviour.' That limitation applies to a Turing machine, but not to every possible machine. Even Turing's biographer Andrew Hodges makes Copeland's list. 'Alan had … discovered something almost … miraculous, the idea of a universal machine that could take over the work of *any* machine.'

'The sooner philosophy and cognitive science are free of this myth the better,' Copeland says. The misunderstanding arose because people forgot to distinguish between what Turing did say, and what he didn't. It's a distinction that Copeland wants to nail down, because it could be the difference between never understanding who and what we are, and making that final, crucial breakthrough in human self-awareness. The truth is, the Turing machine can't take over the work of *any* machine. It certainly can't take over the work of a hypercomputer – as Turing himself made clear.

As we have seen, a Universal Turing Machine is just a very good human mathematician who is able to perform, with great efficiency and speed, any of the tricks available to mathematics. But Turing himself pointed out that this is not the only possibility. In his 1938 PhD thesis he imagined another kind of machine, one that could go beyond the limits of a Universal Turing Machine:

> Let us suppose that we are supplied with some unspecified means of solving problems in number theory; a kind of oracle as it were. We will not go any further into the nature of this oracle than to say it cannot be a machine. With the help of the oracle we could

form a new kind of machine (call them o-machines), having as one of its fundamental processes that of solving a given problem in number theory.

These days, Turing's 'Omega-Machine' is known – to the relatively few computer scientists that are even aware of its existence – as a hypercomputer. It remains a highly speculative, highly controversial area of research. But the potential of hypercomputing makes this long shot well worth our attention.

It is extremely difficult to give a satisfying definition of a hypercomputer. That's because, in every instance where we have identified how one might work, it involves running things for infinite amounts of time, or making perfectly accurate measurements (let's note here for future reference, though, that a Universal Turing Machine also deals in infinite quantities – of time and tape, for example). But don't let that put you off. As a human being, whose brain may turn out to work as a hypercomputer, you have a wondrous capacity for imaginative thinking. So, let's put it to work.

We'll start with the scheme that Jack Copeland and Diane Proudfoot have outlined for creating a hypercomputer that solves the halting problem.

It involves that staple of computing, binary digits: 0s and 1s. When you press a key on your computer keyboard, the processor receives a string of binary digits to process. The letter A is 01000001. From there, you can build up a set of inputs and instructions – a program.

Copeland and Proudfoot define the program the hypercomputer will run very precisely: 'Given an integer that represents a program (for any computer that can be simulated by a universal

Turing machine), output a "1" if the program will terminate or a "0" otherwise.'

Now we look at the machine that performs the hitherto uncomputable task. It is a physical object, and Copeland and Proudfoot suggest a capacitor, a component of electrical circuits. Their capacitor holds a precise amount of electrical charge. That amount can be written as a binary number, which is made up of an infinite string of binary digits.

This is where it gets really eyebrow-raising. That capacitor's charge was chosen because it just so happens to provide a representation of which programs halt, and which don't. Say you want to ask whether a program that can be represented as the number 13 will halt. You find the 13th digit of the capacitor's charge in binary. If it's a 1, that program will halt. If it's a 0, the program will carry on forever. Job done. We have solved the halting problem.

I can almost hear your splutterings. We may have hypothetically solved the halting problem, but only through imagining a completely artificial set of circumstances that we could never hope to realise. All your objections are completely valid. But we have nonetheless just made our minds walk through a scenario in which we can compute the 'uncomputable'.

At the risk of overwhelming your imaginative capacities, let's walk through other seemingly improbable or impossible scenarios that will allow us to hypercompute.

There is, for instance, the Turing machine with an infinite number of symbols inscribed on its tape (and thus boasting an infinite amount of memory) before it begins its operations. Such a change allows it to compute the halting function. Or you can have a network of Turing machines working from the same tape but with a peculiar way of co-ordinating their respective operations on the symbols inscribed on the tape. This allows computations not possible with ordinary Turing machines. Then there's the error-prone hypercomputer; if it prints the wrong symbol in a particular situation, that would allow it to perform non-Turing computations. As Oxford University philosopher Toby Ord has pointed out, we may have inadvertently created hypercomputers like this, but configuring one by accident that did something useful seems a stretch.

We could talk about probabilistic Turing machines. They do not follow a definite path through their program but move to a next step (or a different next step) with a certain probability. Again, they compute the uncomputable, but not in any way that can be prescribed to do something useful. It is possible to see hypercomputing at work in some quantum computing ideas, where a single electron in a hydrogen atom can simultaneously occupy an infinite range of energy levels. There's even a hypercomputer that works on the infinite curvature of spacetime at the centre of a black hole. Infinite-state Turing machines run an infinitely long program, allowing all kinds of computations. Perhaps most pleasing, though, is the hooter-equipped accelerated Turing machine.

The Manchester Mark I machine was equipped with a hooter. It produced what Turing termed 'a steady note' that could be controlled by clever programming to sound any note of any duration. The first big program run by the Mark I announced its halt by playing the British national anthem. Copeland notes that a hoot could signal something far more significant: a computation of the uncomputable.

To understand how, let's define a 'moment'. This is the time it takes for the machine to carry out the first instruction of a program. If its operation were accelerated, so that the machine performed each successive instruction in half the time it took to perform the previous one, it could perform an infinite number of operations within a finite time. This is another way to compute the halting function, and would mean that the hooter would sound within two moments if the machine were to halt. No signal after two moments tells you the machine will run for ever.

There is no shortage of possibilities to explore – there are at least twenty models for hypercomputing. But is any of this really worth our time? As we have seen, ask most computer scientists whether you can have a computer that computes a wider range of things than a Universal Turing Machine and you'll almost certainly get the answer no. That's the point, they'll say: the machine is universal. It covers the whole of mathematically computable functions – you just have to program it correctly and it can do everything that is possible.

The trouble is, that's an extremely anthropocentric view. The Universal Turing Machine, after all, only has the same capabilities as an array of human mathematicians. Copeland puts the self-centredness of this nicely: 'it would – or should – be one of the great astonishments of science if the activity of Mother Nature were never to stray beyond the bounds of Turing-machine computability.'

And let's face it, we already know we naturally give ourselves a very restricted view of Mother Nature's activities. In a paper called 'Machines, Logic and Quantum Physics', David Deutsch, Artur Ekert and Rossella Luppachini make the point that our knowledge of the truth about the universe depends entirely on our knowledge of the laws of physics. And those laws come from our experimental tests. The problem with this is that our tests always take place within the sphere of the physical universe in which human beings work. That means our picture of reality tends to be constrained by our conception of time and sits within just a few dimensions of space. We really shouldn't fall into the trap of thinking that means there is nothing beyond this. After all, we have fallen into that trap before. Take the dominance of Euclidean geometry, for example.

Euclid was a Greek mathematician who lived some time around 300 BC. His textbook, *Elements*, pulled together all facts known about geometry at the time into one volume. Essentially it tells you all you need to know about the properties of things that exist on a flat plane. When you draw a triangle in flat Euclidean space, for instance, the internal angles add up to 180 degrees. *Elements* gives you a definition of parallel lines and other things you might vaguely remember from school. And it wasn't just your school – it was every school, and every academy and university, for thousands of years that followed Euclid; for millennia, Euclidean geometry was really just known as *geometry*.

To give you a flavour of just how dominant Euclidean thinking was, take the case of Carl Friedrich Gauss, the eminent French mathematician. In the early nineteenth century he had the sense that, of the five axioms Euclid had laid out two millennia previously, one was unnecessary. He started working on the problem when he was just fifteen years old, but Gauss never published his results because he was afraid his peers and colleagues would chastise him for questioning the master geometer. It took another decade before two courageous mathematicians called Janos Bolyai and Nicolai Lobachevsky said what others had only thought. Slowly, in a process that took decades to reach acceptance, the fifth axiom was dropped. Gradually, people began to think about the possibility that Euclid didn't have exclusive rights over the shape of things. After more than two thousand years, we broke free and invented other geometries – non-Euclidean geometries.

You almost certainly didn't learn about these at school. That's partly because they can't be made physical in our three-dimensional universe, and are extremely difficult to deal with in the abstract thinking-space inside your head. Hyperbolic geometry, for instance, requires you to deal with parallel lines that gradually curve away from one another and triangles whose internal angles sum to less than 180 degrees. It's useful when you're trying to imagine the shape of the universe, though. Albert Einstein's general theory of relativity required the use of non-Euclidean geometry: the presence of mass and energy changes the geometry of space and time to make it flat and Euclidean, or hyperbolic, or spherical. From 300 BC (and possibly long before) to the beginning of the nineteenth century, no one could imagine anything other than Euclidean geometry. Now we are trying to find out which of the many possible geometrics corresponds to our physical universe. The lesson? Just because we can't wrap our brains around something, it doesn't mean it isn't there.

We are already rediscovering this notion through quantum experiments that take us beyond even Einstein's conceptions of geometry and suggest that the things we thought we understood about the properties of space and time might be completely wrong. Our understanding of the quantum phenomenon of entanglement provides an example. More properly, we should talk about 'non-local correlations' because we are talking about the behaviour of two or more particles that, however widely separated in space they might be, have an instantaneous effect on one another – or as near instantaneous as is possible to measure.

Those measurements have been done. First, you create two entangled photons, then separate them as widely as you can – send them to opposite sides of the universe, if possible. Choose a photon, and measure its 'spin', a quantum mechanical property that can be measured against its orientation in space. Then, as fast as you can, you measure the other particle's spin. What you will find, when you do this again and again, is that there is a pattern in the results of the two measurements – the outcome of the second measurement is dependent on the outcome of the first.

The obvious reason for this would be that some signal is travelling from the first photon to the second, letting it know, so to speak, the result of the measurement. But we have managed to separate these photons widely enough and get the timings of the two measurements close enough that we know that can't be the case. We haven't managed to send particles to opposite sides of the universe, but in 2008 Nicolas Gisin and his colleagues at the University of Geneva sent some to opposite ends of a fibre-optic network. The separation of 18 kilometres was enough to allow Gisin to show that any signal would have to travel between the photons at more than 10,000 times the speed of light. We know that is impossible. The only explanation Gisin can come up with is that the photons exploit some reality that exists beyond space and time.

Going beyond space and time is an emerging theme in quantum theory now, and these innovations at the frontiers of quantum experiments are bringing us a new freedom that should have a bearing on our thoughts about why hypercomputers are impossible. 'Rarely, if ever, has such an important physical claim about the limits of the universe been so widely accepted from such a weak basis of evidence,' says Toby Ord of those who dismiss the possibilities of hypercomputing. 'The inability of some of the best minds of the century to develop ways in which we could build machines with more computational power than Turing machines is not good evidence that this is impossible.'

The bottom line is that we don't know what is impossible. But if we can imagine it, perhaps nature has a way to realise it. At this stage in our development, when our understanding of computing is only a few decades old, it seems presumptuous to dismiss such a possibility – especially given the potential for a hypercomputer to open up the universe of the mind.

Turing was fascinated by the possibility that his computing machines might give us insights into ways we could mimic, or even outperform, human intelligence. That was why he invented the famous 'Turing Test' where a human is asked to hold a conversation with a hidden partner. A machine capable of holding such a conversation without the human becoming aware they were not talking to another human could truly be considered intelligent, Turing suggested.

In his 1948 paper 'Intelligent Machinery', Turing considered how this might be achieved. The paper provides the first effort to describe a machine that works through a system of artificial neurons connected in modifiable ways, mimicking the brain. Turing's neural network could be trained, or educated, by creating or destroying connections between the neurons. With typical brilliance, Turing demonstrated that this system was enough to create a general-purpose computer.

Turing only ever saw his machine in pencil and paper form. In 1958, the year he died, researchers at MIT brought the idea off the drawing board and presented the first real neural network. It has to be admitted that all the intervening years have not given us anything that can mimic the extraordinary abilities of the human brain. But four decades after the first neural network was demonstrated, Hava Siegelmann outlined a version that might.

She called it 'the analog shift map'. It's not a snappy name, but the abstract of her 1995 paper in *Science* is intriguing. The machine 'has computational power beyond the Turing limit … it computes exactly like neural networks and analog machines. This dynamical system is conjectured to describe natural physical phenomena.' And, more to the point – and against everything we thought we knew about hypercomputers – she thinks she might be able to build it.

To understand even vaguely what Siegelmann is trying to build involves getting to grips with some fairly tough concepts. One is number theory, which describes the processes behind mathematics. Another is infinity (and the fact that there are infinitely many infinities). Then there's the basis of it all: neural networks.

A neural network is a computer designed to mimic animal brains. Brain cells – neurons – exist in a vast interconnected web, with each neuron connected to many others. In a neural network the neurons are replaced by simple computer processors that take their input signals, process them in some way and create an output that is fed to another processor. The highly interconnected nature of the network means that the whole thing can do some pretty extraordinary things.

It can learn, for example. A standard computer doesn't learn. It is given an input, and it has a program that acts on the input to produce an output. A neural network, on the other hand, can be configured so that the program changes depending on what data has previously gone through its inputs and outputs, and what the end product of the program has or hasn't achieved. That is how neural networks have been trained to become excellent chess or poker players, to diagnose cancers from scans, to interpret human speech and to drive cars and steer robots.

One of the features of neural networks is that the strengths of connections between the processors can be altered. A processor might have three inputs, for example (a typical neuron in the brain has 10,000), but it doesn't have to give them equal roles, or weighting, in the function it is computing. It might, if it were performing a simple addition of numbers, ignore input number 2, say. If that results in an output or action down the line that is not conducive to the overall goal of the machine, next time it will try dropping, or maybe halving, input number 3 instead. As the machine runs through all the various different possible ways of processing its inputs and outputs, it learns what configurations work well, and what don't. Eventually, like a child learning to ride a bicycle, it finds stable configurations and balances that produce the desired end result.

Now it's time for a little number theory. Standard neural networks describe the weighting of the inputs using rational numbers. These are the integers, or the numbers that can be written as a fraction with integers top and bottom (integers are 'whole numbers': 1, 2, 3 and so on). So our processor might weight its three inputs in the ratio 1, 3/2, 1/2, for example.

Siegelmann, however, showed that you can have a processor that weights its connections in weird ways, and that changes the kinds of computations that a neural network can perform. One trick is to use irrational numbers – numbers, such as pi, that go on and on forever without stopping. That changes the kinds of computations that a neural network can perform. Another is to use chaotic systems, such as the outputs of suitably designed electronic circuits. Their ever-changing, totally unpredictable, uncomputable signals take the neural network beyond Turing computation.

The size of the 'space' in which a standard Turing machine can work – the number of different calculations it can perform – is the same size as the set of natural numbers – the numbers we use for counting. What Siegelmann calls her 'super-Turing' machine has a bigger workspace, the size of which can be described as 2 raised to the power of a Turing machine's workspace. That means, if the Turing machine had a range of 100 calculations, the super-Turing machine's range would be 2^{100}, or 2 followed by 100 zeros. A bigger scope, coupled with the ability to adapt and learn, makes this a serious piece of kit.

And it's being built: funded by the US National Science Foundation, Siegelmann is working with two computing and engineering professors at Missouri State University to turn the idea into a reality. Their goals are fascinating. One is the study of chaotic systems – the kinds of behaviour that almost certainly existed at the Big Bang, as George Gamow suggested, but can't be analysed by human mathematics or (of course) a Turing machine. If Siegelmann's neural network hypercomputer can analyse these systems, it could conceivably change our conception of where we came from. The other main goal is to move towards intelligent machines that exhibit capabilities reminiscent of the human brain. Success in that would tell us a lot about who and what we are.

Can Siegelmann and her colleagues succeed? Most say no, because of the hypercomputer's association with infinities. Irrational numbers are infinitely long, and a processor that relies on anything infinite is surely going to fail – isn't it?

Infinity is an appalling concept. It certainly appalled Georg Cantor's contemporaries, whose refusal to accept his proofs about infinities sent him hurtling into a nervous breakdown.

Cantor, a mathematician working in nineteenth-century Germany, proved there were many different kinds of infinity. There's the straight infinity of integers: however far you count, you can always go one more. Slightly weirdly, the infinity of even numbers is the same size as the infinity of integers, even though the even numbers are contained within a subset of the set of integers. Then, between each one of those integers – between 1 and 2, say – there is an infinite number of real numbers, such as 1.23456789. Cantor showed that this infinity of real numbers is larger than the infinity of integers. Before we drive ourselves into a nervous breakdown, we should perhaps just accept Cantor's proof – now universally accepted – that there is an infinite number of different infinities.

However, though Cantor's proof is mathematically correct, there may be no infinities at all in the real world. Rutgers University mathematician Doron Zeilberger, for instance, says there is no such thing as infinity outside of mathematics. Even inside the subject, the concept is leading us round in circles. We'll get nowhere, he says, until we start working with the idea that there is one final, largest number.

Many physicists, though not quite as radical as Zeilberger, are keen to shed the concept of infinity because it has given us false solutions to many of the problems of cosmology. According to MIT's Max Tegmark, for instance, it is assumptions of a real, physical infinity that created the situation where Alan Guth's inflation is an acceptable theory. Tegmark and others say that physics would describe the universe – and its history – much better if we no longer had infinity as an option in our theories. And that, if it pans out, could undo many of the problems people foresee with schemes such as Siegelmann's: in the real world, measurements of infinite loops or infinite precision may not turn out to be the stumbling blocks that the more traditional computer scientists are predicting. Issues of infinitely precise measurement, for example, are almost certainly a red herring, says Sheffield computer scientist Mike Stannett. Computation, he points out, doesn't actually have to involve measurement – just outputs fed to inputs. And even if we can't build a hypercomputer in this universe, exploring the idea of one is still valuable. We can't build anything that has a non-Euclidean geometry, but understanding its frameworks has still proved remarkably useful in getting to grips with the cosmos. So there is every reason for optimism about hypercomputing. According to Stannett there is 'nothing in mathematics or physics to prevent the implementation of such systems.'

What's more, you might even have one inside your head. Gödel saw his Incompleteness Theorem as an indication that the human brain is something other than a Turing machine, and Turing certainly seemed to suggest that human intelligence requires more than the standard version of computation. A mimic of the brain is certainly where Siegelmann is heading with her neural networks.

As yet we are nowhere near finding out for certain whether the human brain is a hypercomputer, but at least we are now getting some idea of what it is that researchers like Siegelmann are trying to recreate. And maybe the greatest revelation will be confirmation that the universe and the brain work together to pull off a spectacular sleight-of-hand trick on us. In our final excursion to the edge of reality we are going to encounter something that could finally tip you into the abyss. Stop all the clocks: the passage of time is the ultimate illusion.

11

CLOCKING OFF

Time is an illusion

When a man sits with a pretty girl for an hour, it seems like a minute. But let him sit on a hot stove for a minute – and it's longer than any hour. That's relativity.

Albert Einstein

The Hammersmith Hospital, a hotch-potch of buildings put up at various times over the last century, sits to the side of a long, straight road in west London. The original building was constructed in 1912, an elegant orange-red brick edifice fronted with sixty-four huge sash windows and capped by a clock tower. Veer to the right hand side as you approach this building, though, and you'll find a modern, sleek set of additions. And if you time it right, you might just see someone emerge from one of the doors with a huge smile on his or her face.

Let's say it's a man. We'll call him Subject A. He is smiling because this morning, in the confines of this hospital, he has experienced a moment of profound revelation. Though his body has been held almost rigid within the clanking metal tube of a brain-scanning machine, his mind has been released and he has been touching eternity. If you ask him, Subject A will tell you that he has been one with the universe this morning. Press him further, in search of details, and he will shrug and laugh, and try and fail to find words to describe the experience. A typical response might be that he lost all sense of himself, all sense of time, all awareness of space. And then, most likely, he'll repeat himself and tell you that he was simply at one with the universe.

There's no sense in pressing Subject A for more. He is unable to get across to you how wonderful it was to have this experience. He won't be able to tell you tomorrow, or the next day, but he'll still have an inner sense of it. The few experiments of this kind that have been carried out show that the effects can last for more than a year. This, for Subject A, might well have been a life-changing morning.

Let's leave him to wander off, still smiling, to catch a bus home. Then let's push through the door where we met him to find where it all happened. Go down a couple of corridors, through another door or two (these ones are securely locked – you'll need someone with security clearance to let you in) and you'll eventually find the source of Subject A's beatific state. It's a small vial of liquid containing a drug called psilocybin. Here, in the grounds of Hammersmith hospital, Patient A has been tripping on a Class A hallucinogen.

In many ways Hammersmith is the perfect hospital for this research. Though originally built as a workhouse, it was established in 1916 with a £1,000 grant from the Joint War Committee as the Military Orthopaedic Hospital. Its mission was to rehabilitate soldiers who had lost or injured limbs in the fierce tragedy of the First World War. There was no shortage of patients: on the first day of July that year, Britain suffered 60,000 casualties at the opening of the battle of the Somme.

The hospital treated the minds as well as the bodies of its

patients. Most were suffering psychological trauma as a result of their injuries and experiences. Many felt they would never be able to make a useful contribution to society again. The hospital staff did all they could for the physical injuries, then made arrangements for the soldiers to take an active part in hospital life. Some made medical supplies such as surgical splints, operating tables and artificial limbs. Occasionally, based on their own experience, they would come up with improved designs for prostheses. Others would work in local shops, forcing them to get out of the hospital grounds and reacquaint themselves with normality. Many were taught a trade: plumbing, electrical work or tailoring. Robert Jones, the hospital's director, called it 'psychological curative treatment'.

Almost a hundred years later the work goes on. A psychological curative treatment for wounded soldiers is the aim of the psilocybin research carried out today at the Hammersmith Hospital.

A soldier's mind is easily damaged by the trauma of the modern battlefield. The result is known as post-traumatic stress disorder and frequently leaves the sufferer unable to function as a normal human being. One of the few paths open to those who would treat these soldiers is psychotherapy and counselling: the therapist encourages them to talk about their experience and come to terms with it. There is a huge problem with this approach, however. Many of these soldiers are simply incapable of revisiting these worst moments of their lives.

This is where the drugs come in. Studies have begun to show that hallucinogens can leave the brain's chemistry rearranged, changing an overarching mood and altering the way the subject views key moments in their past. What's interesting – to me, here, however – is that the experience is rather like being disconnected, just for a few moments in your life, from the relentless march of time.

Climb into the functional Magnetic Resonance Image (fMRI) scanner at the Hammersmith Hospital, and the operator can watch how blood flows around your brain. When your brain chemistry is altered by a dose of psilocybin, an interesting thing occurs. In the subjects who report the most intense hallucinatory experiences, the blood flow in the posterior singulate cortex and the medial prefrontal cortex is drastically reduced. We know from previous studies that these are the areas in which we encode our sense of self, our knowledge of our spatial and temporal surroundings, and our awareness of how the two interact. Reduce the blood flow here, it seems, and we float away from space and time.

There are other ways to achieve such a blessed state. Nuns in the throes of exceptionally deep prayer have reported a sense of disconnection from time. Buddhist monks say the same. And they are not making exaggerated claims: both groups have rather helpfully conducted prayer sessions inside fMRI scanners. At the moments of greatest spiritual abandonment, blood flow to their posterior singulate cortices and medial prefrontal cortices have been observably diminished. If you take the right drugs or devote yourself enough to religious observance, you can, it seems, make your escape from time.

Maybe that shouldn't be such a surprise. Everything we have learned in the last one hundred years of physics tells us that what we call time is almost certainly just an illusion.

'Absolute, true, and mathematical time, of itself, and from its own nature, flows equably without relation to anything external,' Isaac Newton wrote in the *Principia Mathematica*. 'All motions may be accelerated and retarded, but the flowing of absolute time is not liable to any change.' We can, he said, measure the flow of this absolute time by means of studying motion: it is 'evinced as well from the experiments of the pendulum clock, as by eclipses of the satellites of Jupiter.'

Isaac Newton lived in an era obsessed by clocks. They were central to navigation of the oceans that were opening up to explorers. The Board of Longitude was offering £20,000 to anyone able to create a clock mechanism that could survive the tumult of a sea passage and remain accurate and reliable: this was known to be key to a nation's wealth.

In Newton's universe, time is a simple affair. It flows, and its flow can be measured – it is fundamental to the Newtonian world. There is a present, a future, a past. There is a moment that is 'now' – ever fleeting, always disappearing before the word is even uttered. History is composed of a succession of 'now' moments, definable by events that precede them, and events that follow them. It is a concept that every physicist after Newton has had to unlearn in coming to terms with how things really are.

There are cultures that aren't so enslaved by time: the Uru-Eu-Wau-Wau people of Brazil, for example. Although they can talk in terms of one event happening before or after another, they have no abstract concept of time as something separate, as a background in which things happen. Their language has no word for time as a stand-alone idea. They have no word for 'month' or 'year'.

When the Uru-Eu-Wau-Wau first came into contact with humans from outside their tribe in 1981, more than half of them fell victim to diseases brought in from the 'more developed' world. Conflicts with miners and land-grabbers further decimated their population: they now number fewer than 1,000 individuals. It is likely that their innocence of Newtonian time will also pass away. But let's see whether *you* can shake off your culture and return to the time-blind state enjoyed by the Uru-Eu-Wau-Wau.

'I don't trust these professors who get up and scribble in front of blackboards.' That's how Richard Keating described his motivation for testing Albert Einstein's seemingly outrageous claims.

He had good reason to be sceptical. Keating was a physicist with the US Naval Observatory in Washington DC. He was an expert in the most sensitive timepieces of the day: the Observatory's caesium-powered atomic clocks. A caesium atom has a natural frequency of oscillation – that is, it vibrates at a rate set by the laws of physics. By measuring time as the duration of a certain number of oscillations of a caesium atom, the US Navy had possession of the world's most reliable definition of a second (if you doubt how persistent culture is, note how navies remain, four centuries after the longitude prize, the guardians of timepieces). This provided the US with their standard for time measurements, and it was Keating's job to take that ultra-reliable measurement around the world; the US Navy's master clock stayed in Washington, and Keating would fly an exact copy to locations that required a synchronised signal. Keating knew clocks, and the best ones were astonishingly reliable. Einstein's claim, that this would change if the clock was moved in a certain way, had to be tested before it was accepted.

Einstein's general theory of relativity is often used as shorthand for 'impossibly difficult to understand'. If you are willing to demote time, however, it's not that hard at all. If you can let go, become like the Uru-Eu-Wau-Wau, we can briefly explore the universe according to Einstein.

Though it's not strictly necessary, Einstein started his thinking by considering the properties of light. At the time light was thought of as an electromagnetic wave: freeze it in time and you'd see it is composed of electric and magnetic fields that grow and diminish in intensity as you move along the wave. Aged just sixteen, Einstein imagined exactly this. Rather than freezing the wave, though, he imagined himself running alongside it so that it appeared to be stationary to his eyes. It was a moment of startling revelation, he later said. If you could move with this light, it wouldn't look like light at all. Because its fields appear static, and not repeatedly growing and diminishing, you have changed the fundamental characteristic of the light beam.

One of the great principles of physics is that the laws of physics shouldn't change depending on your circumstances. Galileo used this when trying to explain that the Earth moves round the sun; those who didn't believe it because they didn't experience any motion should try shutting themselves up in a ship, he said. You can't tell if the ship is moving or still, 'so long as the motion is uniform and not fluctuating this way and that.' We have all experienced this at a railway station: a train arriving at the next platform can make you think your train is departing. The sensation – your experience of the physical environment – is the same whether you are moving or not.

Einstein used his teenage insight to formulate a world-changing theory. It started with this premise: if everyone moving steadily through the universe must experience light in exactly the same way, the speed of light must be the same for all observers. In other words, light will always travel at the same speed: the speed of light. It gets complicated in ways to do with geometry and some fairly advanced mathematics, but that's essentially all we need to know.

So the light from your car's headlamps travels at this speed, known as c, even when you are speeding along the highway. The light doesn't go faster just because the bulbs are doing 70 miles per hour. The result, Einstein quickly realised, is rather profound. In May 1905, after a day of discussing physics puzzles with his friend Michele Besso, he announced the Newton-busting paradigm. 'Time cannot be absolutely defined,' he said.

Let me offer a sketch of why this is. We are about to venture into very un-Newtonian territory, but let's start with Newton's insight that we can only measure intervals of time. That, after all, is the only thing a clock can do.

Ignoring the fact that they lived hundreds of years apart, let's put Newton and Einstein together at Cape Canaveral. They both have identical, synchronised clocks. Einstein, though, is taking his clock on a trip through space. He climbs into a rocket, blasts off, and heads through the universe at a speed close to the speed of light. What happens to their clocks defies all your Newtonian intuition.

Both the clocks are composed of atoms, which follow the laws of physics. A fundamental ingredient in many of those laws is the speed of light – when we work out the equations that determine how atoms behave and interact, and how matter and radiation affect each other, the speed of light is always a factor. No wonder, then, that the atoms in Einstein's clock are affected by travelling close to the speed of light relative to Newton's clock. In order to preserve the constant speed of light in their frame of reference, the atoms in Einstein's clock behave differently to those in Newton's clock.

Not that Einstein sees a difference, because the atoms in his body are doing exactly the same thing. The difference is relative – it's Newton, if he could somehow monitor what is going on in the spaceship, who would notice that not only is Einstein's clock running slower than his, but Einstein himself is not ageing as fast.

The interaction between the character of atoms, electromagnetic radiation such as light, the speed of travel relative to Newton and the geometry of the universe required to maintain a constant speed of light reveals something odd about the nature of time: it is flexible, not absolute. The distortions of time involved in any practicable scheme are not big: you have to do very speculative calculations to get an appreciable effect. For instance, if Einstein travelled away from Newton for a hundred years at 0.99 times the speed of light, Einstein would have aged just fourteen years to Newton's one hundred. Einstein's clock would verify this, whether it worked via a mechanical spring, the oscillations of a quartz crystal, pulses of light or radioactive emissions from a lump of uranium. All those processes would be noticeably affected by travelling so close to the speed of light.

Newton would be horrified. It's natural to be horrified. Richard Keating tempered his horror with a physicist's impulse to check the theory – which is why he and Joseph Hafele flew their clocks around the world.

In all the work he had done, Keating had never seen evidence supporting Einstein's prediction. At the 1970 Winter Meeting of the American Physical Society at the Shoreham Hotel in Washington he saw his chance.

Joseph Hafele gave a talk at the meeting about testing Einstein's theory on a jet-plane flight. Hafele, who was based at Washington University, Missouri, was a relatively junior physicist. He had good ideas but no means of funding such an audacious experiment. Keating stepped in. That is how, in 1971, Hafele came to spend $7,000 on two sets of round-the-world plane tickets. He bought four seats under three names: his own, Richard Keating and a Mr Clock. First they would fly a 41.2-hour trip eastward. Then they would fly it west.

You won't be surprised to learn that Mr Clock was one of Keating's atomic clocks (you might be surprised to learn that the airline gave a $400 discount because the clock wasn't going to need feeding). He was big enough to require two seats. There is a photograph of the pair on board an aircraft, with Mr Clock in place and a stewardess checking her watch in the background. They look rather pleased with themselves. And well they might, because the experiment worked perfectly.

The pair flew on Pan American Flight 106 from Washington's Dulles International Airport on 4 October, and spent the expected 41.2 hours in the air. The westward trip around the world began on 13 October and took 48.6 flying hours. On each flight, Keating and Hafele monitored the aircraft's trajectory using information from the flight deck, and kept a keen eye on the clocks' accuracy. They managed to grab just three hours' sleep.

After each trip was over, the travelling clocks were checked against the US Navy's master clock. Both times, there was a discrepancy. On the eastward trip the airborne clocks had lost around 40 billionths of a second. Going west they had gained just under 300 billionths of a second. The clocks went awry on the aircraft for one very simple reason. Wherever you are in the universe, however fast you are travelling, the speed of light never varies – exactly as Einstein had predicted. Newton was unseated.

This is not an isolated result: a raft of atomic clock experiments that followed in Keating and Hafele's footsteps has confirmed the original findings. It's worth stopping for a moment to consider what this means. The emissions of radiation from a caesium atom travelling in a jet airliner are different from the emissions from an identical atom left in one spot on Earth. If that is true for a caesium atom, it is also true for the atoms in your body. Send someone hurtling through space at thousands of miles per hour, and their atoms will go through their processes at a different rate to the people they left behind on Earth. Put simply, they will age at a different rate. Although we can't prove it, we now know, as far as it is possible to know, that space-station astronauts who have spent months or years circling the Earth at enormous speeds (relative to us) have aged a few fractions of a second less than if they had remained Earth-bound.

This has remarkable consequences. The distortion of time – it's known to physicists as 'time dilation' – shatters our concept of cause and effect. First, it is possible for two people to move relative to each other through the universe in such a way that they can both see two events but cannot agree on which one of them happened first. Another casualty is the concept of a universal 'now': one person's present moment is in another person's past, depending on how they are moving relative to one another.

There is no way to make this palatable. Your common sense tells you it can't be true, but common sense is not a useful guide to reality. Time is at best a moveable feast. More likely is that it simply doesn't exist at all, except inside your head. But before we get there, you probably want more evidence. After all, Einstein's relativity is only one of the twin pillars of modern physics. So what does the other one – quantum theory – have to say?

Quantum theory was born and grew up in the first decades of the twentieth century. During the 1950s and 1960s it was a teenager, starting to work out its place in the world, sullenly rebellious to the point where it exasperated many of those who had given it birth, Einstein included. In the late 1970s and through the 1980s it reached a kind of maturity. This was the age of the pocket calculator, the microchip, the first home computers. These days we have begun to understand – and show – that quantum theory is behind everything. Not just everything in electronics, but everything in the whole universe. It is at work in the way your phone operates, it creates your body, it makes stars shine and fills empty space with particles that fizz in and out of existence.

And, somewhat disturbingly, we are discovering that this power behind the cosmic throne has absolutely no need of time.

We have already come across some of the quirks of quantum theory. To appreciate what it has to say about time, we need to look a little more closely.

One is wave-particle duality, which is crucial to the way your mobile phone works. Inside your phone is an array of microprocessors. They are made from tiny bits of silicon whose working parts are the fundamental negatively charged particles known as electrons. I say particles, as if electrons are tiny blobs of negative charge. In truth a more accurate way to describe them would be as a mist. Sometimes you'll see a mist sitting in a valley on a sunny winter's day; you cannot tell where it starts or ends, how close to the ground it sits, or how far up the hillsides you would have to go to escape its clammy touch. This is how we should think of the electrons inside electronics.

It's so useful because they allow electrons to seep, mist-like, through the circuitry. The difference is that this is a mist we can control. Equipped with an understanding of the rules of quantum theory and a means of controlling the voltages at different parts of the circuit, we can turn the flow of charges on and off in ways that, eventually, create powerful computers.

But here's the rub. Some of that flow of charge is, in all normal thinking, impossible. Microchips rely on electrons performing tricks equivalent to you jumping over a 50-metre fence. They manage this because, if they approach the fence, their mistiness, their indefinite physical extent, means that a small part of them is already on the other side. Get all the conditions right, and you can use this to get the whole electron to 'tunnel' through the fence. How long does that take? It's a question without an answer, because quantum theory doesn't deal with time as something real.

It's not that time isn't there at all. You can ask where the electron is at a particular time (well, to be precise, you can ask what is the probability of finding the electron in a certain place after a certain amount of time has elapsed). But this theory – a theory that has passed every experimental test ever thrown at it – has nothing to say about time itself. You can't get any kind of answer to the question 'how long did the electron take to cross the fence?' because the answer is just not something that this theory deals in. Given that this same theory tells us exactly how energy fizzes through empty space and how elements are forged in stars (both confirmed in experiments), its antipathy towards time is suggestive.

Time, it would seem, isn't all that important to the workings of the universe. It is certainly not important to the particles that inhabit it. Take the photon, the particle we associate with light. It simply does not experience time. How could it, when it travels at the speed of light? As we have already seen, the closer you go to the speed of light, the slower time flows. To a photon travelling at the speed of light, time simply does not exist.

The evidence keeps on piling up. In 1967 two renowned physicists, Bryce De Witt and John Wheeler, found a way to bring quantum theory and relativity together, but it involved ditching the notion of time. In October 2013 a group of Italian researchers showed that this kind of universe would still have an appearance of time passing for those within it. They created a physical system involving two photons – the particles of light – and demonstrated that the photons' properties were static when viewed from outside the system, but changed when viewed from within. The explanation comes from the 'observer effect', where quantum properties of particles are affected by being measured. You can have time – but it's not a fundamental. Another argument for the dethronement of time comes from a thought experiment proposed by a physicist at the University of Vienna, Caslav Brukner. His idea is based on a phenomenon that Erwin Schrödinger, one of the theory's chief architects, once tried to disprove: superposition. You might have heard of the 'Schrödinger's cat experiment': it is a reaction to the fact that quantum theory allows particles to be in two places at once, or moving in two directions at once, as long as there is no measurement taking place.

Schrödinger's thought experiment involved placing a cat in a sealed box with a flask of poison gas. The flask can be broken by a hammer whose fall is triggered by the emission of a particle from a radioactive substance. Since the emission of the particle is a quantum event, the equations say that in the absence of any measurement, the particle is both emitted and not yet emitted. That means that, as long as nobody opens the box to look at the cat – and therefore indirectly measure whether the radioactive substance has emitted a particle or not – the cat is both alive and dead.

This seemed ridiculous to Schrödinger, but in quantum theory, the ridiculous is entirely normal. That's clear because in 2003 Anthony Leggett won the Nobel Prize in Physics, in part for creating analogues of Schrödinger's cat in electrical circuits. They weren't dead and alive, but they did have particles travelling through them in two directions at once.

Just as Schrödinger's thought experiment took a while to be realised, Brukner thinks it will take a while to turn his insight into a reality. In 2013 Brukner and his colleagues showed that the equations of quantum theory tell us that particles can be in two moments at once. Not just two places. Not just travelling through space in two different ways. But existing at two separate times. Simultaneously.

It's mind-bending stuff, so let's simplify it a little. Essentially, Einstein can walk into a room and see a message that Newton has left for him. He erases the message, then writes a reply. As Einstein finishes, Newton comes into the room to write the original message. It's a superposition of 'Einstein was in the room before Newton' and 'Newton was in the room before Einstein'. It is impossible, as with relativity, to say who came into the room first.

The result relies on an additional quantum phenomenon: the uncertainty principle, which says it is impossible to extract an unlimited amount of information from a quantum system. That means there is a limit to the precision of any measurement – including a measurement of time. There are, right there in the bowels of our most successful theory of physics, limits to what we can say about anything to do with time. Why? Almost certainly because time isn't a fundamental part of the universe.

In 2008 the Foundational Questions Institute held an essay competition on the nature of time. Some of the world's leading physicists entered and laid out their stall for how we should view time in an age where we know quantum theory as the ultimate arbiter of reality. The entries make fascinating reading. Carlo Rovelli, for instance, suggested that, if we want to unite quantum theory and relativity – the ultimate goal of theoretical physics – 'we must forget the notion of time altogether, and build a quantum theory of gravity where this notion does not appear at all.' Fotini Markopolou was less dismissive. Time, she said, matters – but we have to ditch the notion of space if we want to keep it: 'If we are willing to throw out space, we can keep time and the trade is worth it,' she says.

However, the winner of the competition, Julian Barbour, took no prisoners: 'the quantum universe is static,' he claimed. 'Nothing happens; there is being but no becoming. The flow of time and motion are illusions.' Barbour's essay is well worth reading, though he admits its construction was difficult. 'Writing about the nature of time is a hard task,' Barbour says. 'Unlike the Emperor dressed in nothing, time is nothing dressed in clothes. I can only describe the clothes.' He starts from Newton's assertion, 'Absolute, true, and mathematical time, of itself, and from its own nature, flows equably without relation to anything external', and asserts that 'Newton can be hoist with his own petard if we see what his marvelous laws actually tell us.' The problem is, he says, no one working with classical physics paid proper attention to the nature of time; that's why we are all so emotionally attached to the concept and so traumatised by its strange behaviour in Einstein's relativistic universe and its complete disappearance in the quantum universe. Barbour closes his essay with a quote from Shakespeare:

> When forty winters shall besiege thy brow,
> And dig deep trenches in thy beauty's field,
> Thy youth's proud livery, so gazed on now,
> Will be a tattered weed of small worth held.

'Unlike Newton,' Barbour observed, 'Shakespeare did not attempt to describe time itself, only the differences associated with it.' And differences, according to him, are all we should concern ourselves with. Time, he says, 'should be banished.'

So we know that the fundamental particles that make up our bodies are not in thrall to time. We know there is no universal now, and that perception of future and past, and thus cause and effect, depend on how you are moving through the universe. We know that the massless photons we associate with electromagnetic radiation do not even experience time. Time is clearly not fundamental to the universe. But enough of putting the clothes of physics on something that doesn't exist in order to explain its non-existence, to use Barbour's words. Let's look at our human experiences of time.

Though we are emotionally attached to time, perhaps being released would be a blessing for humans. It certainly worked well for Giacomo Koch's patient. Koch, who works at the University of Rome Tor Vergara, reported the strange case in the journal *Neurology*: a forty-nine-year-old man who had a problem with leaving work at the right time. 'He said he was not able to judge when the working day was over,' Koch reports. It's not clear whether his workplace possessed a clock. Laudably, though, the patient tended to err on the side of caution, and go home early.

Intrigued, Koch assembled a group of healthy people against whom he tested his patient's sense of time. He asked them all to say what length of time had passed between two visual signals. To stop them counting the seconds, they had to read out random numbers that appeared on a screen in front of them. After twenty tests, whose duration varied from 5 seconds to a minute and a half, Koch and his team took a look at their scores.

It's worth pointing out that none of them were terribly good at estimating the shortest intervals of time. At 60 seconds, though, the healthy guessers were almost perfect. Koch's patient thought that was about 40 seconds. By 90 seconds, he was way off beam: his guess was, on average, around the 48-second mark.

Given this underestimation of the passage of time, the patient's early departures from the office seem rather suspect – it would make more sense if he was working unaccountably late into the evening. But let's focus on the lesson. It turned out that the patient had a lesion in the right prefrontal cortex of his brain; a clot in his carotid artery was restricting blood flow. His memory was fine, as was his ability to pay attention to stimuli. The only casualty was his sense of time. Brain damage can wreak havoc on your measure of passing moments. In other words the passage of time is an unquestionably subjective phenomenon. No wonder we report strange phenomena, such as time passing slowly when we are bored, fast when we are busy and in slow motion when we are involved in a dramatic moment.

Everyone has their own story of when time ground almost to a halt. Mine was when a car knocked me from my bicycle. I was twelve years old, and I remember flying through the air after the impact, and thinking how strange it was to be able to think so clearly when airborne. In fact, however, that's almost certainly a false memory: in dramatic moments, time doesn't really slow to the point where you can pay attention to details. We know this because psychologist Chess Stetson once persuaded a group of people to throw themselves off a tower from a height of 46 metres.

The tower was at the Zero Gravity amusement park in Dallas, Texas, but the experience was clearly far from amusing. There's a clue in the title of the research paper they published afterwards: 'Does Time Really Slow Down During a Frightening Event?' One of the researchers involved, David Eagleman, said it was the scariest thing he had ever done.

During their freefall the volunteers read out any numbers they could see on a 'perceptual chronometer' attached to their wrists. This device was a square array of sixty-four red LED lights. A display of the number in LED lights – red on black – was alternated with the illumination pattern inverted: the same number was then displayed black on red. The human eye can see the difference if the alternation happens slowly enough, but at a sharp threshold in speed the two become superimposed and the display looks as if all the lights are on. If time really does slow down and perception is heightened when you are in a frightening, life-threatening situation, Stetson reasoned, the volunteers should be able to see the numbers.

First, Stetson tested the threshold speed at which the volunteers could see the flickering number. Then he gave them a test jump and asked them to estimate how long the fall lasted. With a safety net installed 15 metres above the ground, the fall took 2.49 seconds, but they all overestimated their fall time by roughly one-third. Time, it seems, really was slowing down inside their heads.

Then Stetson made them jump again, this time wearing their perceptual chronometers set to one-third above their threshold flicker speed. If time really was slowing down, they should be able to see a number as they fell.

Stetson ended up with only nineteen data points, because one volunteer kept her eyes closed all the way down. But it made no difference: none of the fallers saw a number. As Stetson put it in the paper reporting the results, 'there is no evidence to support the hypothesis that subjective time as a whole runs in slow motion during frightening events.'

The memory of time is to blame. When people encode memories of an emotionally charged event, they tend to use a 'high-definition' recording. The brain interprets the increase in data as an increase in the time taken for the event. It's a simple mistake.

There are other ways your perception of time is distorted. Disorders such as schizophrenia and drugs such as cocaine and methamphetamine will make time accelerate in your head. Suffering with Parkinson's disease and using marijuana will slow your experience of time. As we have seen, psilocybin can even take you into eternity.

Investigating the tricks time plays on us can help us under-
stand more about how and where in the brain we get our sense of

time passing. At the moment the smart money is on there being a multiplicity of clocks inside your head. In fact the timing processes inside our heads seem to have a lot in common with Einstein's conception of timing in the universe at large.

'A given time interval is registered differently by independent clocks dependent upon the context.' So say Catalin Buhusi and Warren Meck. Unlike Einstein, Buhusi and Meck have a group of trained rats to thank for this insight. Researchers interested in time perception routinely train rats to associate levers with certain time durations. One lever will release a tasty treat if the time between the start and end of an audible signal is 10 seconds or less, say, while another lever releases food when the sound is between 10 and 30 seconds long. Food can be sought at the third lever if the sound lasts between 30 and 90 seconds. It's called the tri-peak procedure, and it seems to show that rats operate three independent clocks, and can start, stop and reset them independently.

Buhusi and Meck's ingenious test involved interrupting the sound signal with an unpredictable and infrequent gap. If the gap came at 15 seconds and lasted 10 seconds, for example, the rats would hit the 10-second lever 30 seconds later than normal, the 30-second lever 20 seconds late and the 90-second lever 10 seconds after they would respond to an uninterrupted signal.

Introducing different gap lengths at different points in the signal gave a set of responses that, when analysed, pointed to a complex set of stops, starts and clock resets induced by the starts, stops and gaps in the signal. The minimum number of independent clocks that would be needed to get such a result, they say, is three. There are almost certainly three clocks in a rat's brain. And there may be more.

We have a fair idea of where some of these clocks reside. Brain scans tell us that the striatum, deep in the forebrain, for instance, is activated when we are trying to assess timings. The striatum is the central message-receiving hub of the basal ganglia, which controls a huge range of the brain's activity. Interestingly, Parkinson's disease affects the striatum's ability to do its job, and affects a sufferer's ability to estimate time intervals.

The prefrontal cortex, the area of the brain damaged in Giacomo Koch's time-challenged patient, also becomes active when we are trying to concentrate on the passage of time. If the part known as the agranular frontal cortex is damaged, a trained rat loses the ability to simultaneously time the duration of two different stimuli, but can still discern the duration of one.

For all this, we are still unsure of how the brain creates time. That it does, though, is surely beyond doubt. Time, according to the biologists who study its creation in the brain, appears to be an emergent property: something that appears in brains of a particular complexity. Which brings us, as we close this chapter, right back to the physicists.

Imagine if time didn't flow. In his beautiful book *Einstein's Dreams*, physicist Alan Lightman takes a moment to do just that. It is, he says, a place for new parents and lovers, where 'the beautiful young daughter with blue eyes and blond hair will never stop smiling the smile she smiles now, will never lose this soft pink glow on her cheeks.' If time stands still, Lightman says, the beloved 'will never fall in love with someone else, will never lose the passion of this instant.' And yet it is not something any of us would choose. Lightman challenges us: is it better to live life, a 'vessel of sadness', or to have an eternity of contentment, 'fixed and frozen, like a butterfly mounted in a case'?

It's a question that physicist James Hartle has pondered from more than a literary perspective. We have to have a flow of time in our heads, he says – otherwise life couldn't have evolved. In 2005 Hartle published an odd paper in the *American Journal of Physics*. It was called 'The Physics of Now', and concerned itself with what would happen if you could make a sentient organism experience time differently.

Hartle has a startling pedigree as a physicist. He co-founded the Institute for Theoretical Physics at the University of California, Santa Barbara. He works with Stephen Hawking on theories of the origin of the universe. He has published many papers on quantum theory with Murray Gell-Mann, a Nobel laureate and originator of the quark model of the nucleus. For all his vast intellect, however, Hartle is not afraid to think about things that don't seem to matter, such as how a frog catches a fly.

To catch a fly, a frog has to take in information about the fly's position and velocity: where it is, and how it is moving. But that information has to be as up-to-date as possible. If it only knows where a fly was 10 seconds ago, the frog will starve. This, Hartle suggests, makes it clear why evolution has equipped us with the time-processing hardware in our brains; to predate on lower organisms requires a sense of past, present and future that the universe, in its cold, relative objectivity, simply doesn't provide. Those creatures whose brains created a sense of time's passing were at a competitive advantage – and so they survived, and the trait flourished. We have a sense of time because we wouldn't be here without one.

To back up his point, Hartle does what most of us couldn't begin to do: he dreams up robots that create time in different ways, and sees the world through their eyes. It's a sorry viewpoint.

There's one that processes information too slowly: the world passes it by. Another has two versions of 'now', 10 seconds apart. This wouldn't survive because its creation of time involves processing more information than a competitor with just one 'now'.

Yet another has no ability to store information. That means it can't have a 'past' from which it can create a simple but useful model of how the world works – we call that model 'experience'. Without a past, you can't learn anything, least of all how a fly moves through the air, so gauging the optimal time to strike involves processing enormous amounts of information. Again, extinction is the inevitable result.

Our brains create time in the only possible way that, given the laws of physics in our universe, permits our survival. That's why a subjective past, present and future is almost certainly a 'cognitive universal', Hartle says. But that doesn't make time real.

Time is subjective. The atoms that make up your body will experience it differently from those moving through the universe in a different way. Your mind experiences it in ways that are shaped by circumstance, by drugs, by your environment, by illness. Newton's absolute time is an illusion.

What is time? 'If no one asks me, I know; but if any Person should require me to tell him, I cannot.' That was St Augustine's response in the year 397. You're in a better position now. Shrug your shoulders and smile at your interrogator. 'It isn't.' That's all you need to say.

EPILOGUE

And the end of all our exploring
Will be to arrive where we started
And know the place for the first time.

T. S. Eliot, Little Gidding, Four Quartets

Jacob Bronowski saw the process of gathering knowledge as an adventure, and it is perhaps adventuring of the most exciting kind. After all, adventuring in the physical world is proving to be a rather finite business.

Just a few decades ago, there seemed to be no end of challenges for the intrepid: reaching the poles, sailing around the world, climbing Everest, scaling the north face of the Eiger, crossing deserts ... these days, few challenges remain. Everest is busier than ever – climbing to its summit is a matter of following a set of rules and procedures, and, ultimately, following a well-maintained set of ropes that are fastened to the mountainside. Just as Everest has shrunk in the adventurer's eyes, so has the range of opportunities open to those who would achieve something extraordinary. Things now have to be done faster, or single-handed, or with ever less help or equipment. These days, explorers tend to discover more about themselves – their motivations and their limits – than the regions through which they are travelling.

It is tempting to see science in the same light – as something that has largely been done. After all, haven't we discovered the structure of DNA, split the atom and traced the history of the universe? In 2013 alone we discovered the Higgs boson, explored the surface of Mars and sequenced the Neanderthal genome. Surely science is running out of challenges?

Not at all. Although more scientists than ever spend their career dotting i's and crossing t's, that is a result of the way science is funded and run, rather than of the opportunities for adventure having run out. For all our progress and discoveries, the unknown remains firmly within sight, teasing us, calling out to us: an unconquered peak that grabs all our attention.

The difficulties that prevent us from scaling the peaks of science are many and varied. Much of the time, the problem is technological. As Freeman Dyson put it, 'New directions in science are launched by new tools much more often than by new concepts.' Often we simply don't have the resources to follow through on the initial presentation of a good idea. When the resources become available, the idea gets explored. Sigmund Freud, for instance, recognised that we should put psychology and psychiatry on a neuroscientific footing a century ago; psychoneuroimmunology is only just beginning, as the tools for reading the states of the brain become available, to mesh with our still nascent understanding of the immune system.

Sometimes, though, we are too busy exploring the implications of one idea to bother with another. Turing's original computing idea is only a few decades old, after all, and we are still marvelling at the world it has created. His other brainchild, hypercomputing, was born at almost the same time, but has had to wait patiently for our attention. The same could be said for epigenetics; there was so much to explore with genetics, is it any wonder we rejected the role of the environment in shaping our inheritance for so long?

There, of course, we have to pin some of the blame on human issues – fashion, fads and petty intolerances. Lamarck's ideas were far from perfect, but they could have been shown far more favour than they were, and this might have accelerated the development of epigenetics as a field. Lamarck's demotion is a complicated one – a result of lack of tools, human foibles and another problem science faces: sometimes we humans simply rush ahead of ourselves and take the easy, but misleading path. The neat story of genetics is being proved too neat, and the story of the Big Bang will also be far less neat when the details are examined.

Other fields take decades to come to fruition simply because they involve years of laborious, painstaking work. The observations of personality and culture beyond the human world were hard-won and had to be rigorous and irrefutable before they could overturn our prejudices. Other fields may never overcome our natural inclinations and become accepted in the mainstream. Researchers have been pushing us towards human–animal chimeras for close to a century, and it's not clear if we are any closer to welcoming this advance. Our discovery of close kinship with other animals may push it yet further away.

Here we encounter one of our biggest problems with science. As it advances, it gives us perspectives far beyond everyday human experience, and the more science we learn, the more insignificant we become. The discovery of a subatomic world where humble particles have godlike properties – hints of omnipresence and freedom from the tyranny of time – are troubling to many. We are similarly dethroned when we see our cultures or personalities mirrored in populations of whales or spiders. But perhaps we shouldn't let this get us down. As we learn of our commonality, we can also celebrate our differences. One is that, for all Jane Goodall's observations, she has never seen a chimp struggling over the philosophy of quantum theory. Though we might never win our battle to understand the reality around us, at least we fight.

The person who pointed out our philosophical superiority over the apes was the astronomer Martin Rees. He has gone further, pointing out that chimps don't even realise there's anything to worry about – they are not aware of quantum theory's existence. Humans, we can be confident, are special, if only in our contemplation of what lies beyond our everyday experience. Though their implications make us feel humble, the very existence of cosmology and quantum theory mark us out as extraordinary biological beings; maybe, in this way, humans have a uniqueness after all. The question facing us now is, how far can we go? Is it possible that understanding some aspects of reality will forever be beyond us? Perhaps – but that doesn't mean we shouldn't keep going, even if progress seems elusive.

It was a scientist – Oxford University's Roger Penrose – who came up with the visual illusion of an eternally rising staircase. He gave the idea to the artist M. C. Escher, who used it to create his masterpiece *Ascending and Descending*. Escher's never-ending quadrangle of steps provides a beautiful analogy for science, which has been walking up its own stairs for many centuries. There is no reason to think we are getting near the top. Neither, though, should we allow ourselves to be fooled into thinking that the climb is futile, that we should just stand still.

As we climb and turn another corner, certain scientific ideas repeatedly come into view, and yet remain obstinately intractable. We have looked at some of them: the quantum roots of biology, the universe as a computer and the nature of consciousness provide good examples of subjects that, for all the research that has been done, remain unresolved. They remain on the edge of uncertainty. And so, therefore, should we.

The edge of uncertainty is not a static line, but a dynamic,

258 AT THE EDGE OF UNCERTAINTY

ever-changing set of answers. Sometimes the answers have needed revision or even rejection, so we keep checking, keep moving, keep stepping forward or sideways – sometimes back. How else could we be? What other way is there for human beings to behave than to push at the boundaries of our knowledge and our existence – even if the act of pushing exposes our ignorance? To quote Richard Feynman, 'it is much more interesting to live not knowing than to have answers that might be wrong.'

In many ways the edge of uncertainty is like the terminator, the line on the Earth's surface where, as the planet turns, darkness gives way to light. The pace of illumination depends where you stand, and what choices you make. Stand at the equator and the terminator will move towards you at a thousand miles an hour. Head towards one of the poles, though, and it moves slower. Once you reach the poles you find that, depending on the time of year, the terminator hardly moves at all. It is possible, at times, to walk ahead of it and keep yourself in day or night, as you choose. If you want to, you can turn and make the sun rise to your west.

Our choices make a difference. It is one of the best traits of humanity that we have chosen to face the coming light, to keep our feet dancing at the edge of uncertainty. We are not necessarily good at science yet; sometimes we head in the wrong direction and light dawns, confusingly, in the western sky. What's more, science speeds up and slows down: the terminator where ignorance meets enlightenment moves forward at an uncertain pace. Much of the time spent doing science is spent in tedium and heartbreak; weeks, months and years can pass without breakthrough, and such times involve painful moments of doubt that a route out of the shrouding ignorance even exists. Progress is almost always hard-earned. Perhaps that is why there is so much joy – sometimes tinged with relief – when the light does come.

Uncertainty, progress, joy and relief; could there be a more concise description of adventure? Gathering new, meaningful knowledge is an endeavour that pushes individuals to their limits. The marvellous thing is that those pioneers open frontiers up for the next generation of explorers. Science might test our resilience, but we have never been defeated. There is always some adventurer who takes up the challenge and moves us forward.

It was great individuals like Albert Einstein and Marie Curie, to name just two, who led us to where we are now. The future will belong to those who bravely – eagerly, even – step into today's uncertainties. Science-fiction writer Ray Bradbury put it beautifully: in science, he said, everything 'begins with romance – the idea that anything is possible.' We have now met some of the current generation of romantics, those to whom the edge of uncertainty sings its siren song. There is space for more, however; this adventure has barely begun.

NOTES

Introduction

3 'Subrahmanyan Chandrasekhar confirmed the suggestion with a mathematical proof': A. Miller, *Empire of the Stars* (Little Brown, 2005); Benjamin Libet tells his story in B. Libet, 'Do We Have Free Will?' in *The Volitional Brain* (Imprint Academic, 1999)

3 'Richard Feynman once said': R. Feynman, *The Feynman Lectures on Physics*, volume I (Addison Wesley Longman, 1964), p. 1–1

3 As George Bernard Shaw put it': G. Bernard Shaw, *Annajanska: The Bolshevik Empress* (1919)

5 'Mach spoke out from the group': J. Bernstein, 'Einstein and the Existence of Atoms', *American Journal of Physics*, vol. 74 (2006) p. 863

6 'The singular literary character Lemony Snicket': Snicket, L., *The Reptile Room* (Egmont, 2001), p. 109

6 'the geologist Eldridge Moores once said': quoted in J. McPhee, *Basin and Range*, (Noonday Press, 1990), p. 214

6 'a mammalian species lasts a million years on average': J. Lawton and R. May, *Extinction Rates*, (Oxford University Press, 1995)

8 'knowledge is "personal and responsible"': J. Bronowski, *The Ascent of Man*, (Little, Brown, 1973), p. 279

8 'as the mystic Deepak Chopra would have us do': see http://store.chopra.com/productinfo.asp?item=73 (but don't buy the book)

1 Triumph of the zombie killers

12 'philosopher David Chalmers coined a phrase': first mentioned at the Tucson Conference on Consciousness in 1994, Chalmers published 'Facing up to the Hard Problem of Consciousness' in the *Journal of Consciousness Studies*, vol. 2 (1995) p. 200. It is available online at http://consc.net/papers/facing.html

15 'Neuroscientist Daniel Bor describes it': D. Bor, *The Ravenous Brain* (Basic Books, 2012), p. 188

15 'That said, it has some heavyweight fans': C. Zimmer, 'Sizing Up Consciousness by Its Bits', *New York Times*, 20 September 2010, available at: http://www.nytimes.com/2010/09/21/science/21consciousness.html?pagewanted=all&_r=0

16 'Dennett published a book with an audacious title': D. Dennett, *Consciousness Explained* (Little, Brown, 1991)

17 'adds up to four hours of blindness': D. Melcher, and C. Colby, 'Trans-Saccadic Perception', *Trends in Cognitive Science*, vol. 12(2008), p. 466

18 'Dennett has performed some stunning (and hugely entertaining) experiments': see, for example, http://www.ted.com/talks/dan_dennett_on_our_consciousness

18 'sometimes you can make an audience blink': E. Dmytryk, *On Film Editing* (Focal Press, 1984)

19 'Together, they made a "Declaration on Consciousness"': see http://fcmconference.org/img/CambridgeDeclarationOnConsciousness.pdf

21 'Adrian Owen's recent discovery': D. Cruse et al., 'Bedside detection of awareness in the vegetative state: a cohort study', *The Lancet*, vol. 378(2011), p. 2088

21 'how Owen put it to *New Scientist* reporter Chelsea Whyte': C. Whyte, 'EEG finds consciousness in people in vegetative state', *New Scientist*, 10 November 2011

22 'R should have been a zombie': C. Philippi, et al., 'Preserved Self-Awareness Following Extensive Bilateral Brain Damage to the Insula, Anterior Cingulate, and Medial Prefrontal Cortices', *PLoS ONE* vol. 7 (8): e38413, doi:10.1371/journal.pone.0038413 (2012)

24 'a paper Turing published in 1950': A. Turing, 'Computing Machinery and Intelligence', *Mind*, vol. LIX, no. 236 (1950), p. 433 , doi: 10.1093/mind/LIX.236.433

25 'An article in *Salon* from 2003 puts it beautifully': J. Sundman, 'Artificial Stupidity', *Salon*, 26 February 2003, http://www.salon.com/2003/02/26/loebner_part_one/

25 '[Turing] wrote a fascinating paper he called 'Intelligent Machinery': best viewed at http://www.turingarchive.org/browse.php/C/11

25 'who called it a "schoolboy's essay" and left it to languish': B. J. Copeland, *Turing: Pioneer of the Information Age* (Oxford University Press, 2012), p. 200

26 'Pain is a curious phenomenon': see http://personal.inet.fi/cool/pentti.haikonen/

27 'When Haikonen beats Experimental Cognitive Robot': you can watch this at http://www.youtube.com/watch?v=48Fh25bXvqk

28 'A 2013 editorial in *Nature*': 'Head Start', *Nature*, vol. 503, November 2013, pp. 5, 7, doi:10.1038/503005a

29 '"We are not the only moral beings."' M. Bekoff and J. Pierce, *Animal Justice* (University of Chicago Press, 2009), p. xv

29 'As Stanford University philosopher Paul Skokowski demonstrates': P. Skokowski, 'I, Zombie', *Conscious Cognition*, vol. 11.1 (2002), p. 1

30 'what Patricia Churchland calls, slightly less elegantly, the "hornswoggle problem"': P. S. Churchland, 'The hornswoggle problem', *Journal of Consciousness Studies* vol. 3 (1996), p. 402

31 '"They're not two things embracing each other,"': S. Blackmore, *Conversations on Consciousness* (OUP, 2005), p. 59

32 'a book called *Animal Personalities*': Samuel D. Gosling and Pranjal H. Mehta, 'Personalities in a Comparative Perspective: What Do Human Psychologists Glean from Animal Personality Studies?', chapter 5 in C. Carere and D. Maestripieri (eds), *Animal Personalities: Behaviour, Physiology and Evolution* (University of Chicago Press, 2013)

2 The crowded pinnacle

33 'Alan Siegel has charted the rise of the riff': A. Siegel, 'How the Song "Seven Nation Army" Conquered the Sports World', deadspin.com, 13 January 2012, http://deadspin.com/5875933/how-the-song-seven-nation-army-conquered-the-sports-world

34 'Luke Martell and Hal Whitehead published a paper': L. Martell and H. Whitehead, 'Culture in Whales and Dolphins', *Behavioral and Brain Sciences*, vol. 24 (2001), p. 309

36 'the older males have been known to visit the younger generation during their first construction efforts': R. Vellenga 'Behavior of the male satin bowerbird at the bower', *Australian Bird Bander*, vol. 1 (1970), p. 3. This work is summarised in J. Diamond, 'Evolution of bowerbirds' bowers: animal origins of the aesthetic sense', *Nature* vol. 297 (1982), pp. 99–102

37 'Take Jane Goodall's experience': J. Goodall in P. Singer and P. Cavalieri, *The Great Ape Project* (St Martin's Press, 1995)

37 'David Hamburg of the Stanford University School of Medicine called it a "once in a generation" endeavour': D. Hamburg in J. Goodall, *In The Shadow of Man* (Phoenix, 1988), p. xi

37 'Stephen Jay Gould called it "one of the great achievements of twentieth century scholarship"': S. Jay Gould in J. Goodall, *In The Shadow of Man* (Phoenix, 1988), p. vii

37 'Masson shared his experiences of derision amongst academics': J. Masson and S. McCarthy, *When Elephants Weep: The Emotional Lives of Animals* (Random House, 2010), p. 12

38 'Charles Darwin was one of the first scientists to stray into this territory': C. Darwin, *The Descent of Man and Selection in Relation to Sex* (D. Appleton, 1896), p. 102

38 'an anthology of readers' tales and anecdotes': G. Romanes, *Animal Intelligence* (D. Appleton, 1882), available at http://www.gutenberg.org/ebooks/40459

39 'L. R. Talbot described his experiences of putting bands on the legs of Georgian birds': L. Talbot, 'Bird-Banding at Thomasville, Georgia, in 1922', available at http://archive.org/stream/jstor-4073430/4073430_djvu.txt

40 'Fruit flies can be categorised into rovers or sitters': J. Mather and D. Logue in C. Carere and D. Maestripieri (eds), *Animal Personalities: Behaviour, Physiology and Evolution* (University of Chicago Press, 2013), p. 18

40 'Jennifer Mather and David Logue's take on the situation': J. Mather and D. Logue in C. Carere and D. Maestripieri (eds), *Animal Personalities: Behaviour, Physiology and Evolution* (University of Chicago Press, 2013), p. 18

41 'gathered themselves a couple of funnel-web spiders, one from the arid deserts of New Mexico and the other from damp woodlands of south-eastern Arizona': S. Riechert and A. Hedrick, 'A test for correlations among fitness linked behavioural traits in the spider *Agelenopsis aperta* (*Araneae, Agelenidae*)', *Animal Behaviour*, vol. 46 (1993), p. 669

43 'In their amusingly titled essay "The Bold and the Spineless"': J. Mather and D. Logue in C. Carere and D. Maestripieri (eds), *Animal Personalities: Behaviour, Physiology and Evolution* (University of Chicago Press, 2013)

43 'shy chipmunks would keep to areas of the reserve where humans are less frequently seen': J. G. A. Martin and D. Réale, 'Temperament, risk assessment and habituation to novelty in eastern chipmunks, *Tamias striatus*', *Animal Behaviour*, vol. 75 (2008), p. 309

44 'A 2004 study of the three-spined stickleback': A. Bell, 'An endocrine disrupter increases growth and risky behavior in threespined stickleback (*Gasterosteus aculeatus*)', *Hormones and Behaviour*, vol. 45 (2004), p. 108

44 'great tits feeding their newborn chicks': R. Cowie and S. Hinsley, 'Feeding ecology of great tits (*Parus major*) and blue tits (*Parus caeruleus*), breeding in suburban gardens', *Journal of Animal Ecology*, vol. 57(1998), p. 611

45 'Birds that had received taurine supplementation as nestlings,': K. Arnold et al., 'Parental prey selection affects risk-taking behaviour and spatial learning in avian offspring' *Proceedings of the Royal Society B*, vol. 274, p. 2563 (2007)

45 'some of the chairs in their waiting room were badly in need of upholstering': M. Friedman and R. Rosenman, *Type A Behaviour and Your Heart*, (Alfred A. Knopf, 1974)

46 'As Sonia Cavigelli and her colleagues put it': S. Cavigelli et al. in C. Carere and D. Maestripieri (eds), *Animal Personalities: Behaviour, Physiology and Evolution* (University of Chicago Press, 2013), p. 479

48 'more than a century and a half of observations of chimpanzees': A. Whiten et al., 'Culture in Chimpanzees', *Nature* vol. 399 (1999), p. 682

49 'laboratory experiments that deliberately induce grief in animals': J. Archer, *The Nature of Grief* (Routledge, 1999)

49 'performed the tests on western scrub jays': T. Iglesias et al., 'Western Scrub-jay Funerals: Cacophonous Aggregations in Response to Dead Conspecifics', *Animal Behaviour*, vol. 84 (2012), p. 1103

49 'King relates the experience of primatologists Christophe Boesch and Hedwige Boesch-Achermann': B. King, *How Animals Grieve* (University of Chicago Press, 2013), p. 83

52 'discovered this behaviour in the South African Kalahari': A. Thornton, 'Variation in Contributions to Teaching by meerkats, *Proceedings of the Royal Society B*, vol. 275 (2008), p. 1745

53 'As Steven Pinker said in *The Language Instinct*': S. Pinker, *The Language Instinct: How The Mind Creates Language* (William Morrow, 1994)

53 'The Linguistic Society of Paris forbade the discussion': see http://www.slp-paris.com/spip.php?article5

53 'The divide was reinforced by Friederich Max Müller': see https://archive.org/details/lecturesonscien07mlgoog

54 'Martin Nowak claims that "language is the most interesting thing to evolve in the last several hundred million years"': M. Nowak, 'Evolution of Language: From Animal Communication to Universal Grammar', Pinkel Lecture (2001), see http://www.ircs.upenn.edu/pinkel/lectures/nowak/

54 'The biologist E. O. Wilson put our dilemma neatly': E. Wilson, 'The Riddle of the
    Human Species', nytimes.com, 24 February 2013, available at http://opinionator.
    blogs.nytimes.com/2013/02/24/the-riddle-of-the-human-species/

3 The chimera era

56 'If you want to retain your admiration for the luminaries of the Enlightenment':
    this story is related in B. Myhre, 'The first recorded blood transfusions: 1656 to
    1668', *Transfusion*, vol. 30(1990), p. 358

59 'only 4 per cent of the eligible UK population donate their blood': see http://www.
    blood.co.uk/giving-blood/

60 'two biomedical researchers filed a patent application': T. Ryan and T. Townes,
    'Production of Human Cells, Tissues, and Organs in Animals', Patent WO
    2000069268 A1

60 '"The State is called upon to produce creatures made in the likeness of the Lord
    and not create monsters that are a mixture of man and ape."' A. Hitler, *Mein
    Kampf*, vol. 2 (Eher Verlag, 1926), ch 2

61 'a Russian historian of science had spent a decade sifting through the evidence':
    K. Rossiianov, 'Beyond Species: Il'ya Ivanov and His Experiments on Cross-
    Breeding Humans with Anthropoid Apes', *Science in Context* vol. 15, (2002), p. 277

65 'Rossiianov was disgusted by what he found out': J. Cohen, 'Zonkeys Are Pretty
    Much My Favorite Animal', *Outside*, 31 July 2007, available at: http://www.
    outsideonline.com/outdoor-adventure/Zonkeys-Are-Pretty-Much-My-Favorite-Animal.
    html

65 'an influential (and highly informative) book on the debate over human-animal
    chimeras': C. MacKellar and D. Albert Jones, *Chimera's Children* (Continuum,
    2012)

65 'David Albert Jones and Calum MacKellar repeat the canard': An email exchange
    between Rossiianov and the author quickly killed any notion this might be true.
    Rossiianov says, 'There is absolutely no evidence to support the claim about
    Stalin's "ape-human soldiers"... I am not even certain that Stalin was aware of
    Ivanov's experiments.'

65 'only traceable to a 2005 *Scotsman* article': C. Stephen and A. Hall, 'Stalin's half-
    man, half-ape super-warriors', *The Scotsman*, 20 December 2005

66 'Your outside is also coated with bacteria': E. Costello et al., *Science*, vol. 336,
    (2012), p. 1255

68 'according to a report from the Academy of Medical Sciences': 'Animals
    Containing Human Material' (July 2011), available at www.acmedsci.ac.uk/
    download.php?file=/images/project/Animalsc.pdf

69 'Hiromitsu Nakaushi and his colleagues in Tokyo and London were celebrating
    an extraordinary success': T. Kobayashi, 'Generation of Rat Pancreas in Mouse by
    Interspecific Blastocyst Injection of Pluripotent Stem Cells', *Cell*, vol. 42 (2010),
    p. 787

69 'researchers from Yokohama City University unveiled the mouse with a human liver': T. Takebe et al., 'Vascularized and Functional Human Liver from an iPSC-derived Organ Bud Transplant', *Nature*, vol. 499 (2013), p. 481

72 'painted the beaks of chickens with luminous T-shirt paint': E. Balaban et al., 'Application of the Quail-chick Chimera System to the Study of Brain Development and Behavior', *Science*, vol. 241 (1988), p. 1339

73 'this could alter the moral status of a monkey': M. Greene et al., 'Moral Issues of Human-Non-Human Primate Neural Grafting', *Science*, vol. 309 (2005), p. 385

73 'Meeting demand for human organs would involve slaughtering a million pigs per year', Editorial: 'Human-animal Hybrids Mean Boom Time for Bioethicists', *New Scientist*, 27 June 2013

73 'Carried out by the Nuffield Council on Bioethics, the report asked for public reactions': 'Animal-to-Human Transplants: the Ethics of Xenotransplantation', available at www.nuffieldbioethics.org/sites/default/files/xenotransplantation.pdf

74 'the BBVA Foundation ... asked 22,500 people a similarly difficult question': 'Second BBVA Foundation International Study on Biotechnology: Attitudes to Stem Cell Research and Hybrid Embryos', available at http://www.fbbva.es/TLFU/dat/international_study_biotechnology_08.pdf

4 The gene genie

77 '"It is now broadly known that an African American man in Harlem is less likely than a man in Bangladesh to survive to the age of 65."': C. Kuzawa and E. Sweet, 'The Embodiment of Race: Health Disparities in the Age of Epigenetics' *American Journal of Human Biology*, vol. 21(2008), p. 2

78 'Diseases such as cardiovascular disease, diabetes and hypertension were all more common in adults that had been born at a weight below the average': D. Barker et al., 'Growth in utero, blood pressure in childhood and adult life, and mortality from cardiovascular disease', *British Medical Journal*, vol. 298 (1989), p. 564

79 '"The more we learn about genes, the more important the environment appears to be."' S. Jones, *The Serpent's Promise* (Little, Brown, 2013), p. 110

79 '"It requires indeed some courage to undertake a labour of such far-reaching extent,"' G. Mendel, Experiments in Plant Hybridization (1865), available at http://www.mendelweb.org/Mendel.html

81 'William Bateson published a book of observations of natural variation': W. Bateson, *Materials for the Study of Variation Treated with Especial Regard to Discontinuity in the Origin of Species*, (Macmillan, 1894)

81 'Matt Ridley wrote in his book *Genome*': M. Ridley, *Genome: The Autobiography of a Species in 23 Chapters* (Fourth Estate, 1999), p. 5

82 'the journal *Nature* called the Internet gateway to the data the "Golden Path"': J. Alfred, 'Golden Path to Genome', *Nature Reviews Genetics* vol. 1 (2000), p. 87

82 'In 2010, in the journal *Cell*': N. Nadeau, 'A Golden Age for Evolutionary Genetics? Genomic Studies of Adaptation in Natural Populations', *Trends in Genetics*, vol. 26(2010), p. 484

82   'According to a 2011 paper in the journal *Science*': M. Przeworski, 'The Golden Age of Human Population Genetics', *Science*, vol. 331 (2011), p. 547

83   'Studies of victims of the Dutch Hunger Winter': L. Schulz, 'The Dutch Hunger Winter and the developmental origins of health and disease', *Proceedings of the National Academies of Science USA*, vol. 107 (2010), p. 16757 (http://www.pnas.org/content/107/39/16757.extract) is a good summary and has many links to relevant studies. The research site http://www.hongerwinter.nl/item.php?id=33&language=EN is another good source

84   'Randy Jirtle and Robert Waterland fed a pregnant agouti mouse': the work was eventually published as R. Waterland and R. Jirtle, 'Transposable Elements: Targets for Early Nutritional Effects on Epigenetic Gene Regulation' *Molecular and Cell Biology*, vol. 23 (2003), p. 5293

86   'Megan Hitchins found a case of a hereditary colorectal cancer marked out by epigenetics': M. Hitchins, 'Inheritance of a Cancer-Associated MLH1 Germ-Line Epimutation', *New England Journal of Medicine*, vol. 356 (2007), p. 697

86   'silenced genes responsible for the onset of Prader-Willi and Angelman syndrome had been methylated': K. Buiting, 'Epimutations in Prader-Willi and Angelman Syndromes: A Molecular Study of 136 Patients with an Imprinting Defect', *American Journal of Human Genetics*, vol. 72 (2003), p. 571

87   'The most definitive proof came via a project known as the "MRC Vitamin Study"': a history of the study is available at http://www.mrc.ac.uk/Achievementsimpact/Storiesofimpact/Folicacid/index.htm

88   'Charles Darwin called him a "celebrated naturalist"': C. Darwin, *On The Origin of Species*, p. xiii, (John Murray, 1859)

89   '"Heaven forfend me from Lamarck nonsense of "a tendency to progression", Darwin wrote': see http://www.darwinproject.ac.uk/letter/entry-729

91   'a team led by epigeneticist Douglas Ruden': D. Ruden, 'Waddington's Widget: Hsp90 and the Inheritance of Acquired Characters', *Seminars in Cell and Developmental Biology*, vol. 14 (2003), p. 301

93   'Bacarelli's team analysed the blood of sixty-seven plant workers': A. Bacarelli et al., 'DNA Methylation Differences in Exposed Workers and Nearby Residents of the Ma Ta Phut Industrial Estate, Rayong, Thailand', *International Journal of Epidemiology*, vol. 41 (2012), p. 1753

93   'air pollution seems to result in methylation of the gene responsible for nitric oxide production': C. Breton, 'Particulate Matter, DNA Methylation in Nitric Oxide Synthase, and Childhood Respiratory Disease', *Environmental Health Perspectives*, vol. 120 (2012), p. 1320

93   'methyl groups can exert an influence on thousands of genes': R. Schmitz, 'Transgenerational Epigenetic Instability Is a Source of Novel Methylation Variants', *Science*, vol. 334 (2011), p. 369

94   'approximately 1 per cent of the changes get through the erasure process unscathed': J. Hackett, 'Germline DNA Demethylation Dynamics and Imprint Erasure Through 5-Hydroxymethylcytosine', *Science*, vol. 339 (2012), p. 448

95   'Bygren's starting point was a sample of 99 people born in 1905': L. Bygren et al., 'Longevity determined by ancestors' overnutrition during their slow growth period', *Acta Biotheoretica*, vol. 49 (2001), p. 53

95   'If boys start smoking before the age of eleven': M. Pembrey, 'Sex-specific, Male-line Transgenerational Responses in Humans', *European Journal of Human Genetics*, vol. 14 (2006), p. 159

96   'According to historians Adrian Desmond and James Moore', A. Desmond and J. Moore, *Darwin's Sacred Cause: How a Hatred of Slavery Shaped Darwin's Views on Human Evolution*, (Houghton Mifflin Harcourt, 2009)

96   'His journal, published in 1845, records his encounters with "heart-sickening atrocities"': A. Desmond, 'Darwin the abolitionist', *Prospect*, 28 February 2009

98   'white women on low incomes have babies that are, on average, 200 grams heavier than those born to black women on equivalent incomes': R. Goldenberg et al, 'Medical, psychosocial, and behavioral risk factors do not explain the increased risk for low birth weight among black women', *American Journal of Obstetrics and Gynaecology*, vol. 175 (2006), p. 1317

98   'A Johns Hopkins hospital survey of their data from 1897 to 1935': D. Costa, 'Race and Pregnancy Outcomes in the Twentieth Century: A Long-term Comparison', *Journal of Economic History*, vol. 64 (2004), p. 1056

98   'A 2006 study of intergenerational birth weights': C. Colen, 'Maternal Upward Socioeconomic Mobility and Black–White Disparities in Infant Birthweight', *American Journal of Public Health*, vol. 96 (2006), p. 2032

98   'A study of 18,000 babies, carried out across the US in 1967': A. Naylor and N. Myrianthopoulos, 'The relation of ethnic and selected socio-economic factors to human birth-weight', *Annals of Human Genetics*, vol. 31 (1967), p. 71

99   'in 1997 two Chicago physicians compared the birth weights of over 90,000 infants': R. David and J. Collins Jr, 'Differing Birth Weight among Infants of U.S.-Born Blacks, African-Born Blacks, and U.S.-Born Whites', *New England Journal of Medicine*, vol. 337 (1997), p. 1209

5 Different for girls

101   'increased the percentage of children with measles immunity from 73 per cent to 100 per cent': S. Plotkin et al. (eds), *Vaccines* (Elsevier Health Sciences, 2012), p. 368

101   'reports coming from Haiti, Senegal, Gambia and Guinea-Bissau had made it clear that there was a serious problem': K. Knudsen, 'Child Mortality Following Standard, Medium or High Titre Measles Immunization In West Africa', *International Journal of Epidemiology*, vol. 25 (1996), p. 665

101   'contrition certainly pervades Neal Halsey's account of his involvement in the story': N. Halsey, 'Increased Mortality After High Titer Measles Vaccines: Too Much of a Good Thing', *Paediatric Infectious Disease Journal*, vol. 12 (1993), p. 462

102   'As a 2010 *Nature* editorial stated': 'Putting gender on the agenda', *Nature*, vol. 465 (2010), p. 665

103 'One of the earliest reliable papers on sex differences in health came in 1959': H. Franke and J. Schröder, *Deutsche Medizinische Wochenschrift*, vol. 84 (1959), p. 653

103 'A report on this paper in *New Scientist*': 'Sex Clues in Disease', *New Scientist*, 17 September 1959, p. 471

103 'a 1965 discovery by Thomas Washburn and colleagues': T. Washburn et al., 'Sex Differences in Susceptibility to Infections', *Paediatrics*, vol. 35 (1965), p. 57

104 'women did indeed have higher levels of immunoglobin M': M. Butterworth et al., 'Influence of Sex on Immunoglobulin Levels', *Nature*, vol. 214 (1967), p. 1224

104 'a group of researchers from London's Royal Free Hospital went one better': K. Rhodes, 'Immunoglobulins and the X-chromosome', *British Medical Journal*, vol. 3 (1969), p. 439

104 'Triple X syndrome, as it is known, is associated with a raft of difficulties': M. Otter, 'Triple X Syndrome: A Review of the Literature', *European Journal of Human Genetics*, vol. 18 (2010), p. 265

105 'In 2004, for instance, it was more than three men to one woman': A. Kim, 'Sex bias in trials and treatment must end', *Nature*, vol. 465 (2010), p. 688

105 'Baggio has assembled a frightening roster of diseases': G. Baggio et al., 'Gender Medicine: A Task for the Third Millennium', *Clinical Chemistry and Laboratory Medicine*, vol. 51 (2013), p. 713

106 'In a blistering comment piece in *Nature*': A. Kim, 'Sex Bias in Trials and Treatment Must End', *Nature*, vol. 465 (2010), p. 688

106 'Erika Check Hayden has written it even more succinctly': E. Check Hayden, 'Sex Bias Blights Drug Studies', *Nature*, vol. 464 (2010), p. 332

107 'ibuprofen worked to reduce pain for the male volunteers, but not for the females': J. Walker and J. Carmody, 'Experimental Pain in Healthy Human Subjects: Gender Differences in Nociception and in Response to Ibuprofen', *Anaesthesia and Analgesia*, vol. 86 (1998), p. 1257

107 'Carmody carried out a study that would account for this': B. Butcher and J. Carmody, 'Sex Differences in Analgesic Response to Ibuprofen are Influenced by Expectancy', *European Journal of Pain*, vol. 16 (2012), p. 1005

108 'In a broad swathe of trials, 400 micrograms of ibuprofen has been found to be effective in both men and women': J. Barden et al., 'Ibuprofen 400 mg is Effective in Women, and Women are Well Represented in Trials,' *BMC Anesthesiology* vol. 2 (2002), p. 6

108 'affect men and women in different ways': I. Campesi et al., 'Sex and Gender Aspects in Anesthetics and Pain Medication', *Handbook of Experimental Pharmacology*, vol. 214 (2012), p. 265

108 'women found the kappa-opiods rather effective': R. Gear et al., 'Gender Difference in Analgesic Response to the Kappa-Opioid Pentazocine', *Neuroscience Letters*, vol. 205 (1996), p. 207

109 'A classic study': D. Hoffmann and A. Tarzian, 'The Girl Who Cried Pain: A Bias Against Women in the Treatment of Pain', *Journal of Law, Medicine and Ethics*, vol. 29 (2001), p. 13

109 'As Dalhousie University's Anita Unruh put it': A. Unruh, 'Gender Variations in Clinical Pain Experience', *Pain*, vol. 65 (1996), p.123

110 'For the women, there is more going on in the parts of the brain associated with attention and emotion': L. Melton, 'His Pain, Her Pain', *New Scientist*, 19 January 2002

110 'a rat will move around more at certain points in its oestrus cycle': G. H. Wang, 'The Relation Between "Spontaneous" Activity and Oestrous Cycle in the White Rat', *Comparative Psychology Monographs*, vol. 2 (1923), p. 27

110 'It's not easy being a man with osteoporosis, for instance': G. Baggio et al., 'Gender Medicine: A Task for the Third Millennium', *Clinical Chemistry and Laboratory Medicine*, vol. 51 (2013), p. 713

111 'tests on the bisphosphonate drugs were done on women aged sixty-five to eighty': T. Järvinen et al., 'The True Cost of Pharmacological Disease Prevention', *British Medical Journal*, vol. 342 (2011), p. 2175

112 'There are nine major categories of birth defects, and only in the nervous system defects are girls worse off': J. Lary, 'Sex Differences in the Prevalence of Human Birth Defects: A Population-based Study', *Teratology*, vol. 64 (2001), p. 237

113 'The pattern of infection was simple, the researchers discovered': P. Aaby et al., 'Overcrowding and Intensive Exposure as Determinants of Measles Mortality', *American Journal of Epidemiology*, vol. 120 (1984), p. 49

113 'it was all about the intensity of the infection': P. Aaby, 'Malnutrition and Overcrowding/Intensive Exposure in Severe Measles Infection: Review of Community Studies', *Review of Infectious Disease*, vol. 10 (1988), p. 478

114 'the same phenomenon in data on the 1885 measles outbreak in Sunderland, England': P. Aaby et al., 'Severe Measles in Sunderland, 1885: A European-African Comparison of Causes of Severe Infection', *International Journal of Epidemiology*, vol. 15 (1986), p. 101

116 'As the London Science Museum website points out': see http://www.sciencemuseum.org.uk/broughttolife/people/jamesphipps.aspx

116 'he hadn't done the experiment on enough people and was thus offering only circumstantial evidence': see http://www.ncbi.nlm.nih.gov/pmc/articles/PMC1044113/?page=1

116 'Näslund compared the survival rates of children who had been vaccinated with the rates of those who hadn't': C. Näslund, 'Resultats des experiences de vaccination par le BCG poursuivies dans le Norrbotten (Suède)', *Vaccination Preventative de Tuberculose, Rapports et Documents* (Institut Pasteur, Paris) (1932)

117 'BCG reduces deaths caused by things other than accidents or TB by 25 per cent': P. Aaby et al., 'Randomized Trial of BCG Vaccination at Birth to Low-birth-weight Children: Beneficial Nonspecific Effects in the Neonatal Period?', *Journal of Infectious Disease*, vol. 204 (2011), p. 245

117 'smallpox and BCG vaccines can reduce susceptibility to lymphoma and leukaemia and asthma': M. Villumsen et al., 'Risk of Lymphoma and Leukaemia after Bacille Calmette-Guérin and Smallpox Vaccination: A Danish Case-cohort Study', *Vaccine*, vol. 27 (2009), p. 6950; P. Bager et al., Smallpox Vaccination and

Risk of Allergy and Asthma, *Journal of Allergy and Clinical Immunology*, vol. 111 (2003), p. 1227

117 'has been used to fight multiple sclerosis and type 1 diabetes': D. Faustman, 'Proof-of-Concept, Randomized, Controlled Clinical Trial of Bacillus-Calmette-Guérin for Treatment of Long-Term Type 1 Diabetes', *PLoS ONE* 7(8): e41756

118 'for unknown reasons, females have a naturally stronger type 2 bias': S. Huber and B. Pfaeffle, 'Differential Th1 and Th2 Cell Responses in Male and Female BALB/c Mice Infected with Coxsackievirus Group B type 3', *Journal of Virology*, vol. 68 (1994), p. 5126

118 'Males hardly switch on any genes, while nearly all the genes are being switched on in females': K. Flanagan, 'Transcriptional Profiling Technology for Studying Vaccine Responses: An Untapped Goldmine', *Methods*, vol. 60, (2013), p. 269; also personal conversation with the author

118 'produced different amounts of protein, depending upon the sex of the mouse', X. Yang, 'Tissue-specific Expression and Regulation of Sexually Dimorphic Genes in Mice', *Genome Research*, vol. 16 (2006), p. 995

6 Will to live

120 'In 1985, William Keatinge, a physiologist based in London, carried out a remarkable experiment': W. Keatinge et al., 'Exceptional Case of Survival in Cold Water', *British Medical Journal*, vol. 292 (1986), p. 171

121 '*New Scientist* was not so polite': S. Young, 'In Cold Blood', *New Scientist*, 22 January 1987, p. 40

121 'the average damaged ship takes at least fifteen to thirty minutes to sink': *NOAA Diving Manual* (Best Publishing, 1977), p. 61

122 'Helena Karppinen published an extraordinary study': H. Karppinen et al., 'Will-to-live and Survival in a 10-year Follow-up Among Older People', *Age and Ageing*, vol. 41 (2012), p. 789

124 'Metalnikoff's results were published in 1926': sources on the history of psychoneuroimmunology are R. Ader, 'Historical Perspectives on Psychoneuroimmunology', available at: www.fundacionsalud.org.ar/images/GetAttachment.doc_ADER_2.pdf; E. Kandel, 'A New Intellectual Framework for Psychiatry', *American Journal of Psychiatry*, vol. 155 (1998), p. 457 ; J. Jones, 'Freud's Authority', *Bookslut*, October 2007, available at: http://www.bookslut.com/psychoslut/2007_10_011780.php; E. Kandel, Nobel Prize Autobiography, available at http://www.nobelprize.org/nobel_prizes/medicine/laureates/2000/kandel-bio.html

128 'Kandel says he "stuck it to the Austrians"': 'Three Q's', *Science*, vol. 320 (2008), p. 1269

129 'The storage of memories is a "very molecular process"' see: http://www.nobelprize.org/nobel_prizes/medicine/laureates/2000/kandel-speech.html

132 'Soili Lehto assembled nineteen depressive patients and put them through a PET scanner': S. Lehto et al., 'Changes in Midbrain Serotonin Transporter Availability

in Atypically Depressed Subjects after One Year of Psychotherapy', *Progress in Neuropsychopharmacological and Biological Psychiatry*, vol. 32 (2008), p. 229

132 'neuropsychiatrist Lewis Baxter compared psychotherapy with the effect of the antidepressant fluoxetine': L. Baxter et al., 'Caudate Glucose Metabolic Rate Changes with Both Drug and Behavior Therapy for Obsessive-Compulsive Disorder', *Archives of General Psychiatry*, vol. 49 (1992), p. 681

133 'Hasse Karlsson, professor of psychiatry at the University of Helsinki looked at twenty studies of brain changes induced by psychotherapy': H. Karlsson, 'How psychotherapy changes the brain', *Psychiatric Times*, vol. 28 (2011), no. 8

133 'As Kandel puts it, "psychotherapy is a biological treatment, a brain therapy"': E. Kandel, 'The New Science of Mind and the Future of Knowledge', *Neuron*, vol. 80 (2013), p. 546

134 'he published a hugely influential paper on "copycat suicides"': for a great long read on Phillips' life and work, see J. De Wyze, 'Why do they die?' *San Diego Reader*, 31 March 2005, available at http://www.sandiegoreader.com/news/2005/mar/31/why-do-they-die/

134 'Phillips first examined the birthdates and deathdates of 1,251 "notable" Americans': D. Phillips, 'Deathday and Birthday: An Unexpected Connection' in J. Tanur (ed.) *Statistics: A Guide to the Unknown* (Holden-Day, 1972)

135 'This was published in *The Lancet* in 1988': D. Phillips and E. King, 'Death Takes a Holiday: Mortality Surrounding Major Social Occasions', *The Lancet*, vol. 32 (1988), p. 728

136 'In a 2001 study, Harvard University's Laura Kubzansky': L. Kubzansky et al., 'Is the Glass Half Empty or Half Full? A Prospective Study of Optimism and Coronary Heart Disease in the Normative Aging Study', *Psychosomatic Medicine*, vol. 63 (2001), p. 910

136 'an article written to burst the bubble of the exaggerated claims of psychological effects on cancer': J. Coyne and H. Tennen, 'Positive Psychology in Cancer Care: Bad Science, Exaggerated Claims, and Unproven Medicine', *Annals of Behavioral Medicine*, vol. 39 (2010), p. 16

138 'the wound took 24 per cent longer to heal': J. Kiecolt-Glaser et al., 'Slowing of Wound Healing by Psychological Stress', *The Lancet*, vol. 346 (1995), p. 1194

138 'the leukocytes in the stressed carers' blood produced less infection-fighting interleukin-1-beta when put in contact with an antigen': R. Glaser and J. Kiecolt-Glaser, 'Stress-induced Immune Dysfunction: Implications for Health', *Nature Reviews Immunology*, vol. 5 (2005), p. 243

138 'A report written for the National Institutes of Health in 1994': J. Achterberg et al., 'Mind-Body Interventions' in *Expanding Medical Horizons: A Report to the National Institutes of Health* (Diane, 1994)

138 'a man whose wife has just died had a 25 per cent higher chance of dying in those twelve years': K. Helsing, 'Factors Associated with Mortality after Widowhood', *American Journal of Public Health*, vol. 71 (1981), p. 802

138 'A meta-analysis – essentially, a study of studies – published in 2011': J. Moon et al., 'Widowhood and Mortality: A Meta-Analysis', *PLoS ONE*, vol. 6, e23465 (2011)

138 'the immune system responses of twenty-six recently bereaved widows and widowers': R. Bartrop, 'Depressed Lymphocyte Function after Bereavement', *The Lancet*, vol. 1 (1977), p. 834

139 'it is not just infection you should fear if bereaved': M. Jones et al., 'The Long-term Impact of Bereavement upon Spouse Health: A 10-year Follow-up', *Acta Neuropsychiatrica*, vol. 22 (2010), p. 212

139 'In 2010, a study enumerated the dangers of loneliness': J. Holt-Lunstad et al., 'Social Relationships and Mortality Risk: A Meta-analytic Review'. *PLoS Med* vol. 7, e1000316 (2010)

139 'being chronically lonely was associated with being almost twice as likely to die': Y. Luo, 'Loneliness, Health, and Mortality in Old Age: A National Longitudinal Study', *Social Science & Medicine*, vol. 74 (2012), p. 907

140 '"Sickness is a normal response to infection"': R. Dantzer et al., 'From Inflammation to Sickness and Depression: When the Immune System Subjugates the Brain', *Nature Reviews Neuroscience*, vol. 9 (2008), p. 45

141 'As Glaswegian researchers Rajeev Krishnadas and Jonathan Cavanagh put it': R. Krishnadas and J. Cavanagh, 'Depression: An Inflammatory Illness?', *Journal of Neurological and Neurosurgical Psychiatry*, vol. 83 (2012), p. 495

141 'A 2003 study published in the *Journal of Psychosomatic Medicine*': M. Irwin, 'Effects of a Behavioral Intervention, Tai Chi Chih, on Varicella-zoster Virus Specific Immunity and Health Functioning in Older Adults', *Journal of Psychosomatic Medicine*, vol. 65 (2003), p. 824

141 'putting them in a room to talk to each other for an hour once a week': D. Spiegel, 'Mind Matters in Cancer Survival', *Psycho-Oncology*, vol. 21 (2012), p. 588

142 '"a new level of cooperation between neurology and psychiatry"': E. Kandel, 'The New Science of Mind and the Future of Knowledge', *Neuron*, vol. 80 (2013), p. 546

7 Correlations in creation

144 'the physicist Erwin Schrödinger published a book called *What is Life?*': E. Schrödinger, *What is Life?* (Cambridge University Press, 1967)

147 'The chemists couldn't quite get the formulation right': the history of this subject is entertainingly explored out in S. Pain, 'Stench Warfare', *New Scientist*, 7 July 2001

148 'his huge, hilarious and utterly absorbing book *Perfumes*': L. Turin and T. Sanchez, *Perfumes* (Profile, 2008)

149 'In the early 1970s, several researchers pointed out that dozens of molecules smell the same': see L. Turin and F. Yoshii, 'Structure-odor Relations: A Modern Perspective', in R. Doty (ed) *Handbook of Olfaction and Gustation* (Marcel-Decker, 2003), p. 275

150 'It came from a researcher called Malcolm Dyson': G. M. Dyson, 'The Scientific Basis of Odour', *Chemistry and Industry*, vol. 57 (1938), p. 647

151 'As Turin says, it's surely "too marvellous to be a coincidence"': M. O'Hare, 'A Nose for Controversy', *New Scientist*, 18 November 2006

152 'In 2011, Turin and his colleagues published a paper describing how they had selected the flies': M. Franco et al., 'Molecular Vibration-sensing Component in Drosophila melanogaster Olfaction', *Proceedings of the National Academies of Sciences USA*, vol. 108 (2011), p. 3797

154 'Sara Maitland's beautiful and startling story *Moss Witch*': S. Maitland, *Moss Witch and Other Stories* (Comma Press, 2013)

155 'Scholes wears Prada shoes and a porn-star shirt': see www.chem.utoronto.ca/staff/SCHOLES/people.html

155 'The group's 2010 *Nature* paper': E. Collini et al., *Nature*, vol. 463 (2010), p. 644

157 'It is hard to measure the efficiency of a plant's conversion of light to electricity': see, for a discussion, J. Bolton and D. Hall, 'The Maximum Efficiency of Photosynthesis', *Photochemistry and Photobiology*, vol. 53 (1991), p. 545

158 'the one that Wiltschko himself references is in 1968': the Wiltschkos are retired now, but a good summary of their work is in R. Wiltschko and W. Wiltschko, 'Avian Magnetic Compass: Its Functional Properties and Physical Basis', *Current Zoology*, vol. 56 (2010), p. 265

159 'The 2010 meeting was to discuss not a problem, but an intriguing possibility': W. Wiltschko et al., 'The Mechanism of the Avian Magnetic Compass', *Procedia Chemistry* vol. 3 (2011), p. 276

160 'experiments have revealed that the radical pair lasts for an extraordinarily long time': E. Gauger, 'Sustained Quantum Coherence and Entanglement in the Avian Compass', *Physical Review Letters*, vol. 106 (2011), p. 40503

161 'the birds might see the magnetic field as a distortion of their normal vision': M. Stoneham et al., 'A New Model for Magnetoreception', *Biophysical Journal*, vol. 102 (2012), p. 961

8 The reality machine

164 'The laws of Nature are information about information': V. Vedral, *Decoding Reality: The Universe As Quantum Information* (OUP Oxford, 2010)

165 '"Information is Physical" was the battlecry of a man': you can read about Rolf Landauer's life and contributions to physics at www.nasonline.org/publications/biographical...pdfs/landauer-rolf.pdf

166 'Stephen Hawking conducted an imaginary experiment': a good introduction to Hawking radiation is at http://math.ucr.edu/home/baez/physics/Relativity/BlackHoles/hawking.html

167 '"Stephen, as we all know, is by far the most stubborn and infuriating person in the universe"': L. Susskind, 'Twenty Years of Debate with Stephen' in G. Gibbons et al. (eds), *The Future of Theoretical Physics and Cosmology* (Cambridge University Press 2003)

167 'Erhard was "America's first self-help guru"': you can get a good sense of Erhard from L. Kellaway, 'Lunch with the FT: Werner Erhard' available at http://www.ft.com/cms/s/2/feb214a8-8f88-11e1-98b1-00144feab49a.html

169 'In a 2008 interview with the *LA Times*': J. Johnson Jr, 'Black Holes and Scientific Standoffs', *LA Times*, 26 July 2008, available at http://articles.latimes.com/2008/jul/26/science/sci-susskind26

169 'Maldacena made a big breakthrough': J. Maldacena, 'The Large N Limit of Superconformal Field Theories and Supergravity', *Advances in Theoretical and Mathematical Physics*, vol. 2 (1998), p. 231, available at arXiv:hep-th/9711200

170 'In July 2004 Hawking sent a note to the organiser of the 17th International Conference on General Relativity and Gravitation': see http://www.theguardian.com/science/2004/jul/15/spaceexploration.highereducation

172 'Hogan did some calculations and found exactly where the pixels of the universe would start to show up': C. Hogan, 'Now Broadcasting in Planck Definition' (2013), available at http://arxiv.org/abs/1307.2283

173 'Douglas Adams portrays our planet as part of a huge computer': D. Adams, *The Hitchhiker's Guide to the Galaxy* (Pan, 1979)

173 'an Isaac Asimov story, *The Last Question*, in which humans seek to reverse the inexorable increase of disorder': I. Asimov, *The Last Question* (Columbia, 1956)

174 'In Fredkin's conception, there is no such thing as space, or time': a wonderful introduction to Fredkin is R. Wright, 'Did the Universe Just Happen?', *The Atlantic Monthly*, April 1988

176 'his astonishing book *Programming the Universe*': S. Lloyd, *Programming the Universe* (Alfred Knopf, 2006)

176 'Pagels had described a premonition of his death': H. Pagels, *The Cosmic Code: Quantum Physics as the Language of Nature* (Simon and Schuster, 1982)

177 'Lloyd came up with the first design for a quantum computer that really could be built': S. Lloyd, 'A Potentially Realizable Quantum Computer', *Science* vol. 261 (1993), p. 1569

178 'a factoring challenge put out by the security firm RSA': the solution announcement is at https://listserv.nodak.edu/cgi-bin/wa.exe?A2=NMBRTHRY;612109bb.1207

178 'a computer operating by the laws of quantum physics could cut through the factoring problem like a hot knife through butter': a guide to Shor's algorithm can be found at http://www.princeton.edu/~achaney/tmve/wiki100k/docs/Shor_s_algorithm.html

179 'Seth Lloyd has calculated just how powerful a computer the universe is': S. Lloyd, 'Ultimate Physical Limits to Computation', *Nature* vol. 406 (2000), p. 1047

181 'David Lindley's excellent book on quantum theory: *Where Does the Weirdness Go?*': D. Lindley, *Where Does the Weirdness Go?: Why Quantum Mechanics is Strange, But Not as Strange as You Think* (Basic, 1996)

182 'Their predictions matched perfectly with their experiments': L. Hackermüller et al., 'Decoherence of Matter Waves by Thermal Emission of Radiation', *Nature*, vol. 427 (2004), p. 711

183 'versions of it have been done in laboratories': V. Jaques, 'Experimental Realization of Wheeler's Delayed-Choice Gedanken Experiment', *Science*, vol. 315 (2007), p. 966

183 'we become participators in the processes of the universe, as Wheeler put it':
J. Wheeler, 'Information, Physics, Quantum: The Search for Links', *Proceedings
of the 3rd International Symposium on the Foundations of Quantum Mechanics*
(Physical Society of Japan, 1989), p. 354

183 '"We are a way for the cosmos to know itself"': C. Sagan, *Cosmos*, Episode 1 (PBS)

184 '"If your theory is found to be against the second law of thermodynamics I can
give you no hope,"': A. Eddington, *The Nature of the Physical World: Gifford
Lectures, 1927* (Cambridge University Press, 2012)

184 'information theory, quantum theory and thermodynamics seem to be
intertwined': M. Müller et al., 'Unifying Typical Entanglement and Coin Tossing:
On Randomization in Probabilistic Theories', *Communications in Mathematical
Physics*, vol. 316 (2012), p. 441

185 'God, Vedral says, is a thermodynamicist': V. Vedral, 'In from the Cold', *New
Scientist*, 15 October 2012, p. 32

9 Complicating the cosmos

187 'Michael Turner of the University of Chicago, went on the record': M. Brooks,
'Inflation Deflated: The Big Bang's Toughest Test', *New Scientist*, 6 June 2008

188 'Georgii grew up in Odessa, on the north-western shores of the Black Sea':
Karl Hufbauer, *George Gamow: 1904–1968 – a biographical memoir* (National
Academy of Sciences, 2009)

190 'Richard Feynman once described turbulence as "the most important unsolved
problem of classical physics"': R. Feynman, *The Feynman Lectures on Physics*
(Addison-Wesley, 1964)

190 '"the primordial gas was in a state of large-scale irregular motion, that is, it was
in a turbulent state."': G. Gamow, 'On the Formation of Protogalaxies in the
Turbulent Primordial Gas', *Proceedings of the National Academy of Sciences USA*,
vol. 40 (1954), p. 480

191 'wingtips of aircraft always create vortices, but here you can see them form': see
http://www.youtube.com/watch?v=QZkggFzAtEc

193 'published a paper focusing in on the "anomalous alignment" and "uncanny"
correlations': K. Land and J. Magueijo, 'The Axis of Evil', *Physical Review Letters*,
vol. 95 (2005), p. 71301

194 'With the Planck data, the Axis of Evil was suddenly causing a real problem':
J. Aron and L. Grossman, 'Cosmic Rise and Fall Writ in the Sky', *New Scientist*,
27 March 2013

195 'When they finally published their tale of galaxy clusters streaming through the
universe': A. Kashlinsky et al., 'A New Measurement of the Bulk Flow of X-ray
Luminous Clusters of Galaxies', *The Astrophysical Journal Letters*, vol. 712 (2010),
p. L81

196 'we have also found a supergiant structure': R. Clowes et al., 'A Structure in
the Early Universe at z ~ 1.3 that Exceeds the Homogeneity Scale of the R-W
Concordance Cosmology', *Monthly Notices of the Royal Astronomical Society*,
vol. 429 (2013), p. 2910

197 'the laws of physics might change over space': J. Webb et al., 'Indications of a Spatial Variation of the Fine Structure Constant', *Physical Review Letters*, vol. 107 (2011), p. 191101

197 'Murphy summed it up beautifully': M. Brooks, 'Constant Change: Are There No Universal Laws?', *New Scientist*, 25 October 2010

198 'a reviewer's report on a paper by Michael Longo': A. Ananthaswamy, 'Original Spin: Was the Universe Born Whirling?' *New Scientist*, 17 October 2011

198 'Look along this line and you'll find more galaxies spinning one way than the other': M. Longo, 'Detection of a Dipole in the Handedness of Spiral Galaxies with Redshifts z~0.04z~0.04', *Physics Letters B*, vol. 699 (2011). It's worth noting that George Gamow asked if the universe was spinning in 1946: G. Gamow, 'Is the universe rotating?', *Nature*, vol. 158 (1946), p. 549

200 'François and Monique Spite were carrying out observations using the Canada France-Hawaii telescope': A review of the Spites' careers is in R. Cayrel, 'From Lithium to Uranium: Elemental Tracers of Early Cosmic Evolution', *Proceedings of the IAU Symposium, no. 228* (International Astronomical Union, 2005), available at journals.cambridge.org/article_S1743921305005144

201 'there is too much Lithium-6': M. Ashland, 'Lithium Isotopic Abundances in Metal-poor Halo Stars', *Astrophysical Journal*, vol. 644 (2006), p. 229

201 'Accounting for them will make cosmology "too bloody complicated"': J. Aron, 'Largest Structure Challenges Einstein's Smooth Cosmos', *New Scientist*, 11 January 2013

205 'when it reported its results in March 2013, they put inflation in some serious difficulties': Z. Merali, 'Higgs Data Could Spell Trouble for Leading Big Bang Theory', *Nature*, 16 April 2013

205 'a paper written by the team running the Planck telescope mission': P. Ade et al., 'Planck 2013 Results. XXIV. Constraints on Primordial Non-Gaussianity', http:// arxiv.org/abs/1303.5084

205 'a quest to make this problem public': A. Ijjas, 'Inflationary Paradigm in Trouble after Planck 2013', *Physics Letters B*, vol. 723 (2013), p. 261

208 'The quote actually comes from Nicolaus Copernicus': N. Copernicus, *De Revolutionibus Orbium Celestial* (1543)

10 Hypercomputing

209 'a handy wall chart about chemical terrorism': see http://www.health.ny.gov/ environmental/emergency/chemical_terrorism/poster.htm

209 'Around 18 to 20 per cent of men cannot smell cyanide': R. Kirk and N. Stenhouse, 'Ability to Smell Solutions of Potassium Cyanide', *Nature*, vol. 171 (1953), p. 698

210 'an essay published in 2012': J. Copeland, 'Turing Suicide Verdict in Doubt' (Oxford University Press, 2012), available at http://fds.oup.com/www.oup.co.uk/pdf/ general/popularscience/jackcopelandjune2.pdf

210 'According to Turing's biographer Andrew Hodges': see http://www.turing.org.uk/ bio/part8.html

212 'the programmer's handbook for the Mark II Manchester computer': available at http://www.computer50.org/kgill/mark1/progman.html

213 'This method fell out of favour a few centuries ago for purely technological reasons': A. Noyes, *Rethinking School Mathematics* (SAGE, 2007), p. 74

214 'what the forthcoming Windows 98 operating system would be able to do': see https://www.youtube.com/watch?v=UjZQGRATIwA

215 'Bertrand Russell's reaction was to declare himself puzzled': B. Russell, *Letter to Leon Henkin, 1 April 1963* (Russell papers, McMaster University, Hamilton, Ontario)

215 'compared Russell's efforts to understand Gödel to a dog staring at a blank screen': D. Hofstadter, *I Am a Strange Loop* (Basic Books, 2007)

215 'The proof of the halting problem': A. Turing, 'On Computable Numbers, With an Application to the Entscheidungsproblem', *Proceedings of the London Mathematical Society*, series 2, vol. 42 (1936–7), p. 230

216 'when two apparent truths stand in conflict with one another': J. Good in 'The Men Who Cracked Enigma', *Heroes of World War II* (History Channel, 2004)

216 'Copeland has compiled a rogues gallery of researchers': J. Copeland, 'The Church-Turing Thesis', in E. Zalta (ed.), *The Stanford Encyclopaedia of Philosophy*, p. 1095, available at http://plato.stanford.edu/entries/church-turing/

217 'In his 1938 PhD thesis': A. Turing, 'Systems of Logic Based on Ordinals' PhD thesis, Princeton University (1938)

218 'a hypercomputer that solves the halting problem': B. J. Copeland and D. Proudfoot, 'Alan Turing's Forgotten Ideas in Computer Science', *Scientific American*, April 1999, p. 99

220 'As Oxford University philosopher Toby Ord has pointed out': T. Ord, 'The many forms of hypercomputation,' *Applied Mathematics and Computation*, vol. 178 (2006), p. 143

220 'a hoot could signal something far more significant: a computation of the uncomputable': B.J. Copeland, 'Even Turing Machines Can Compute Uncomputable Functions', in C. Calude et al., *Unconventional Models of Computation* (Springer-Verlag, 1998)

221 'our knowledge of the truth about the universe depends entirely on our knowledge of the laws of physics': D. Deutsch et al., 'Machines, Logic and Quantum Physics', available at http://arxiv.org/abs/math/9911150

222 'Gauss never published his results': there is a nice discussion of this at http://www.jeremychapman.info/cms/one-too-many-the-role-of-euclid%E2%80%99s-fifth-postulate-in-the-development-of-non-euclidean-geometries

224 'Nicolas Gisin and his colleagues at the University of Geneva sent some to opposite ends of a fiber-optic network': D. Salart et al., 'Testing the speed of "spooky action" at a distance', *Nature*, vol. 454 (2008), p. 861

224 'photons exploit some reality that exists beyond space and time': J-D. Bancal et al., 'Quantum Non-locality Based on Finite-speed Causal Influences Leads to Superluminal Signalling', *Nature Physics*, vol. 8 (2012), p. 867

224 'Says Toby Ord': T. Ord, 'Hypercomputation: Computing More than the Turing Machine', available at http://arxiv.org/abs/math/0209332

225 'In his 1948 paper called "Intelligent Machinery"': A. Turing, 'Intelligent Machinery', a report for the National Physical Laboratory (1948), available at: http://www.alanturing.net/intelligent_machinery/

225 'the abstract of her 1995 paper in *Science* is intriguing': H. Siegelmann, 'Computation beyond the Turing limit', *Science*, vol. 268 (1995), p. 545

229 'no such thing as infinity outside of mathematics': for an interesting discussion of the problems with infinity, see A. Gefter, 'The Infinity Illusion', *New Scientist*, 17 August 2013, p. 32

229 'Computation, he points out, doesn't actually have to involve measurement': M. Stannett, 'The case for hypercomputation', *Applied Mathematics and Computation*, vol. 178 (2006), p. 8

229 'Gödel saw his Incompleteness Theorem as an indication that the human brain is something other than a Turing machine': W. Sieg, 'Gödel on computability' *Philosophia Mathematica*, vol. 14 (2006), p. 189

11 Clocking off

232 'The hospital treated the minds as well as the bodies of its patients': for some of the history, see http://ezitis.myzen.co.uk/hammersmith.html and http://archive.org/st ream/1a3reveille00galsuoft/1a3reveille00galsuoft_djvu.txt

233 'hallucinogens can leave the brain's chemistry rearranged': see, for example, R. Carhart-Harris et al., 'Implications for Psychedelic-assisted Psychotherapy: Functional Magnetic Resonance Imaging Study with Psilocybin', *British Journal of Psychiatry*, vol. 200 (2012), p. 238 and R. McKie, 'Magic Mushrooms' Psychedelic Ingredient Could Help Treat People With Severe Depression', *The Observer*, 7 April 2013, available at http://www.theguardian.com/science/2013/apr/07/magic-mushrooms-treat-depression

234 'Reduce the blood flow here, it seems, and we float away from space and time': interview with author and R. Carhart-Harris et al., 'Neural Correlates of the Psychedelic State as Determined by fMRI Studies with Psilocybin', *Proceedings of the National Academy of Sciences USA*, vol. 109 (2012), p. 2138

234 'There are other ways to achieve such a blessed state': D. Biello, 'Searching for God in the Brain', *Scientific American Mind*, October/November 2007, p. 38

235 'no abstract concept of time as something separate': C. Sinha et al., 'When Time Is Not Space: The Social and Linguistic Construction of Time Intervals and Temporal Event Relations in an Amazonian Culture', *Language and Cognition*, vol. 3 (2011), p. 137

236 '"I don't trust these professors who get up and scribble in front of blackboards"': R. Hafele in 'Time Travel', NOVA (PBS, 1999), transcript at http://www.pbs.org/wgbh/nova/transcripts/2612time.html

238 '"Time cannot be absolutely defined," he said': A. Einstein, 'How I Created the Theory of Relativity' (translation by Yoshimara A. Ono) *Physics Today*, vol. 35 (1982), p. 45

243 'a group of Italian researchers showed that this kind of universe would still have an appearance of time passing': E. Moreva et al., 'Time from Quantum Entanglement: An Experimental Illustration', available at http://arxiv.org/abs/1310.4691

243 'thought experiment proposed by a physicist at the University of Vienna, Caslav Brukner': O. Oreshkov et al., 'Quantum Correlations with No Causal Order', *Nature Communications*, vol. 3 (2012), p. 1092

245 'In 2008, the Foundational Questions Institute held an essay competition': all the mentioned essays are available at: http://fqxi.org/community/essay/winners/2008.1

247 'reported the strange case in the journal *Neurology*': G. Koch et al., 'Selective Deficit of Time Perception in a Patient With Right Prefrontal Cortex Lesion': *Neurology*, vol. 59 (2002), p. 1658

248 'There's a clue in the title of the research paper' C. Stetson et al., 'Does Time Really Slow Down During a Frightening Event?' *PLoS ONE*, vol. 2, e1295 (2007)

250 'Buhusi and Meck have a group of trained rats to thank for this insight': C. Buhusi and W. Meck, 'What Makes Us Tick? Functional and Neural Mechanisms of Interval Timing', *Nature Reviews Neuroscience* vol. 6 (2005), p. 755

251 'Parkinson's disease affects the striatum's ability to do its job, and affects a sufferer's ability to estimate time intervals.' M. Pastor et al., 'Time Estimation and Reproduction is Abnormal in Parkinson's Disease' *Brain*, vol. 115 (1992), p. 211

251 'Alan Lightman takes a moment to do just that' A. Lightman, *Einstein's Dreams* (Pantheon, 1993)

251 'We have to have a flow of time in our heads, he says – otherwise life couldn't have evolved': J. Hartle, 'The Physics of Now', *American Journal of Physics*, vol. 73 (2005), p. 101

Epilogue

255 'As Freeman Dyson put it': F. Dyson, *Imagined Worlds* (Harvard University Press, 1998)

257 'The person who pointed out our philosophical superiority over the apes': H. Blake, 'Royal Astronomer: "Aliens May Be Staring Us in the Face"', *Daily Telegraph*, 22 February 2010

258 'To quote Richard Feynman': the quote comes from C. Sykes (ed.), *No Ordinary Genius: The Illustrated Richard Feynman* (W. W. Norton, 1994), p. 239

259 'Science-fiction writer Ray Bradbury put it beautifully': E. Hoover, 'Mars Teems With Myths, Symbology' *Los Angeles Times*, 9 August 1976, p. B1

INDEX